WITHDRAWN

Complex Numbers

CHAPMAN AND HALL
MATHEMATICS SERIES

Edited by Professor R. Brown,
Head of Department of Pure Mathematics,
University College of North Wales, Bangor,
and Dr Michael Dempster,
Lecturer in Industrial Mathematics
and Fellow of Balliol College, Oxford

A Preliminary Course in Analysis
R. M. F. Moss *and* G. T. Roberts

Elementary Differential Equations
R. L. E. Schwarzenberger

A First Course on Complex Functions
G. J. O. Jameson

Rings, Modules and Linear Algebra
B. Hartley *and* T. O. Hawkes

Regular Algebra and Finite Machines
J. H. Conway

Complex Numbers

A study in algebraic structure

W. H. COCKCROFT

G. F. Grant Professor of Pure Mathematics
University of Hull

LONDON
CHAPMAN AND HALL

First published 1972
by Chapman and Hall Ltd
11 New Fetter Lane, London EC4P 4EE
Filmset by Keyspools Ltd, Golborne, Lancs
Printed in Great Britain by
Redwood Press Limited
Trowbridge, Wiltshire

SBN 412 11510 7
LC 73-155360

Published in the U.S.A.
by Halsted Press, a Division
of John Wiley & Sons, Inc.
New York

Preface

This book is the outcome of a lecture course which was intended to introduce first year undergraduates to some of the ideas of modern abstract algebra in a context with which they had some familiarity, namely the complex numbers.

The lectures given were often discursive in character and in consequence the book has some of this quality. Throughout I have tried to recognize that learning mathematics involves a process in which one moves from the imprecise and intuitive to the precise and rigorous. In particular the algebra with which the book is concerned has associated with it a whole series of geometrically intuitive ideas, which the student cannot, and indeed in my opinion should not, ignore. To what extent these geometrical ideas should be formalized in any rigorous fashion in dealing with the algebra of complex numbers is possibly debatable. I have chosen to encourage as much informal use of appropriate geometrical ideas as possible at each stage, reserving a rigorous approach for the algebraic work proper. This may or may not please pure algebraists, but is done in recognition of the fact that students of the subject do not come to any new ideas with empty minds. I have tried to build on their understanding of past work, rather than ignore it.

My thanks are due to my many colleagues who have discussed the subject matter and the approach with me, especially to J. D. Cooke and J. R. Dennett, who read the manuscript during preparation.

Preface

Contents

Introduction

Our first contact with formal mathematics comes when we meet the natural numbers 1, 2, 3, ..., and their rules for addition and multiplication. As our early mathematical education proceeds we are introduced to more kinds of numbers; we meet in turn the *integers*, the *rational* numbers, and to some extent the *real* numbers. We are thus led, rightly, to believe that mathematics is very much concerned with 'numbers' and their rules of manipulation.

Our classroom experience with the various kinds of numbers reflects our practical needs. Thus our work with the natural numbers first arises from our need to be able to count. The (positive) rational numbers, i.e. the ratios m/n of natural numbers, m, n, arise from our subsequent need to divide things up, and then lead us to be able to measure with greater accuracy. We meet negative numbers when we need to be able to count and measure not only 'forwards' but also 'backwards'.

More mathematically, of course, we can view the introduction of rational numbers as enabling us to divide one natural number by another, whether or not the result is again a natural number. Similarly the introduction of negative numbers may be viewed as a recognition of a need to be able always to define the difference between any two positive numbers a, b, whether a is greater or less than b.

Continuing in mathematical vein, the study of real numbers becomes vitally important when we need to cope with limiting processes.

Numbers other than positive real numbers have not always been acceptable to mathematicians. Thus in the case of negative numbers, the ability to subtract one number from another has not always meant an acceptance of negative numbers as such. For example, knowledge in the sixteenth century of formulae such as

$$(a-b)(c-d) = ac-bc-ad+bd$$

implied an ability to manipulate expressions involving negative signs but certainly did not imply a use of the formula except in the case $a > b, c > d$. Again, negative solutions of algebraic equations were at one time regarded as non-admissible. It would appear that something in the nature of a 'real' mathematical interpretation of purely negative numbers was required before they became entirely acceptable to mathematicians. This was provided by the now completely familiar 'real number line', extended both to the left and the right of the origin 0, so that numbers in this context can be regarded as *directed*, positively to the right of 0

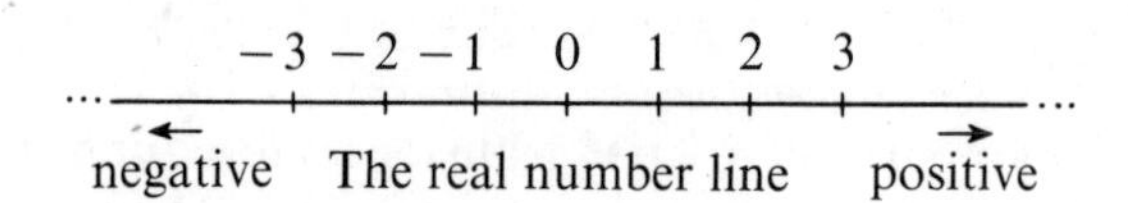

The real number line

and negatively to the left. Linked as it became with scientific usage (temperature scales provide a typical example of such use), this geometrical model quickly led to the recognition of negative numbers as numbers in their own right with entirely acceptable rules of manipulation.

If negative real numbers have not always been acceptable as solutions of algebraic equations, it is not surprising that mathematicians have not always accepted that equations such as $x^2+1 = 0$ could have solutions which were meaningful. It is one thing to write $x^2 = -1$, whence $x = \pm\sqrt{(-1)}$, and another to accept that the square root of -1 should be considered to be a 'number'. Formal manipulations such as those involved in the factorization

$$40 = (5+\sqrt{(-15)})(5-\sqrt{(-15)})$$

were made, again in the sixteenth century, using the ordinary rule

$$(a-b)(a+b) = a^2-b^2$$

and writing $-(\sqrt{(-15)}(\sqrt{(-15)}) = -(-15) = 15$. But such manipulation did not imply a belief that the expression $5+\sqrt{(-15)}$ should be thought to have meaning. Indeed, in attempting to solve quadratic equations

$$px^2+qx+r = 0,$$

where p, q, r are real, and $q^2 < 4pr$, we ourselves find expressions which can be written in the form $a+b\sqrt{(-1)}$, where a and b are real. We can then formulate, using ordinary algebraic rules, the rules for addition and multiplication of such expressions:

$$(a+b\sqrt{(-1)})+(c+d\sqrt{(-1)}) = (a+c)+(b+d)\sqrt{(-1)},$$

$$(a+b\sqrt{(-1)})(c+d\sqrt{(-1)}) = (ac-bd)+(bc+ad)\sqrt{(-1)},$$

simply writing -1 for $(\sqrt{(-1)})(\sqrt{(-1)})$. However, we need, as earlier mathematicians needed, some suitable concrete interpretation of such expressions if we are to find them acceptable as 'numbers'. Otherwise we must contrast them with the real numbers, judging them to be 'imaginary', as did our predecessors.

The interpretation which led historically to the acceptance of these *complex numbers* $a+b\sqrt{(-1)}$ generalizes the geometrical interpretation which led to the acceptance of negative numbers, and like that interpretation is still found satisfactory. In place of the real line we take the real plane and associate with each 'number' $a+b\sqrt{(-1)}$ the point in the plane with co-ordinates (a, b). The real numbers 'a' are then taken as x-co-ordinates and the real multiples 'b' of $\sqrt{(-1)}$ are taken as y-co-ordinates. In this context we usually refer to the x-axis in the plane as the *real* axis and the y-axis as the *imaginary* axis.

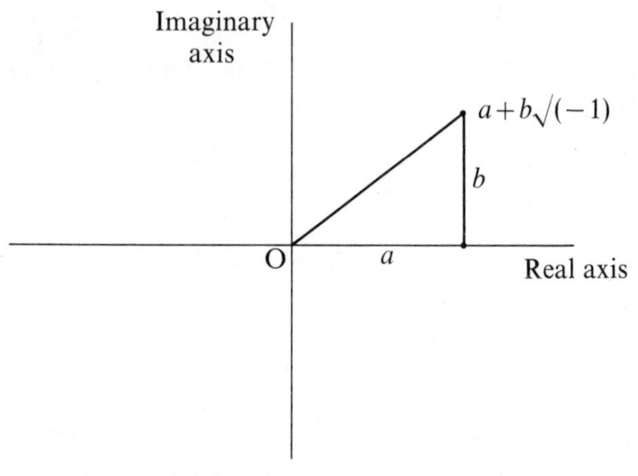

Diagrammatic representation of $a+b\sqrt{(-1)}$

Of course it is not sufficient to interpret only the expressions $a+b\sqrt{(-1)}$. We must also make meaningful the algebraic manipulation of such expressions, i.e. their addition and multiplication. Before doing this in general, it is worthwhile considering the particular case of multiplication of a real number by $\sqrt{(-1)}$, since it was the geometrical interpretation of this operation which led eighteenth-century mathematicians to accept the 'reality' of the complex numbers.

Notice then that formal multiplication of a real number x by $\sqrt{(-1)}$ converts it into $x\sqrt{(-1)}$. On our diagram the corresponding point $(x, 0)$ on the real axis is therefore transformed by an anticlockwise rotation through a right angle about the origin into the point $(0, x)$ on the imaginary axis. A further multiplication by $\sqrt{(-1)}$ yields $x(\sqrt{(-1)})^2 = -x$, and the point $(0, x)$ is transformed, again by an anti-clockwise rotation about the origin, into the point $(-x, 0)$ on the real axis. Interpreted in this way the equation $(\sqrt{(-1)})^2 = -1$ takes on a reality in rotational terms which led our predecessors to accept the complex numbers as meaningful, just as the directing of numbers on the real number line had earlier led to the acceptance of negative numbers.

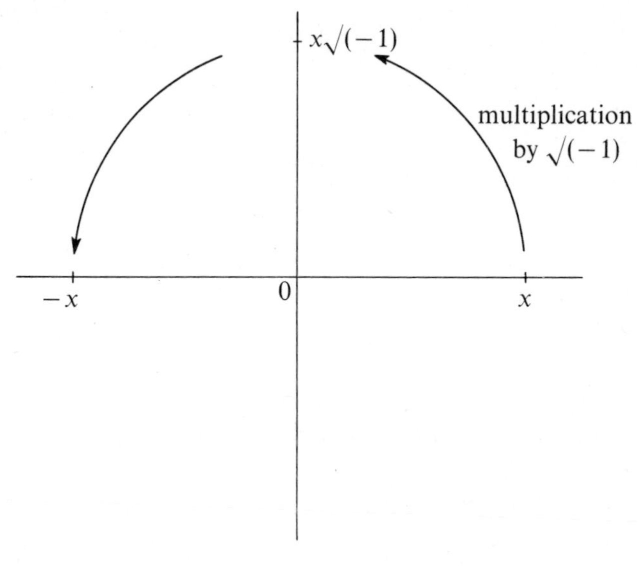

$\sqrt{(-1)}$ as a rotation through $\frac{\pi}{2}$

Turning to the general geometric interpretation of addition and multiplication, let us at the same time follow standard practice and write i for $\sqrt{(-1)}$. Addition of complex numbers then takes the form

$$(a+bi)+(c+di) = (a+c)+(b+d)i.$$

If we associate with $a+bi$ the point P with co-ordinates (a, b) and with $c+di$ the point Q with co-ordinates (c, d), then with the sum of $a+bi$ and $c+di$ we must associate the point R with co-ordinates $(a+c,\ b+d)$. The point R completes the parallelogram OPRQ. Addition of complex numbers can in consequence be interpreted as an addition of displacement vectors, for we can associate with $a+bi$ not only the point P(a, b) but also the displacement vector $\overrightarrow{\mathrm{OP}}$, from the origin 0 to the point P, and similarly with $c+di$ the

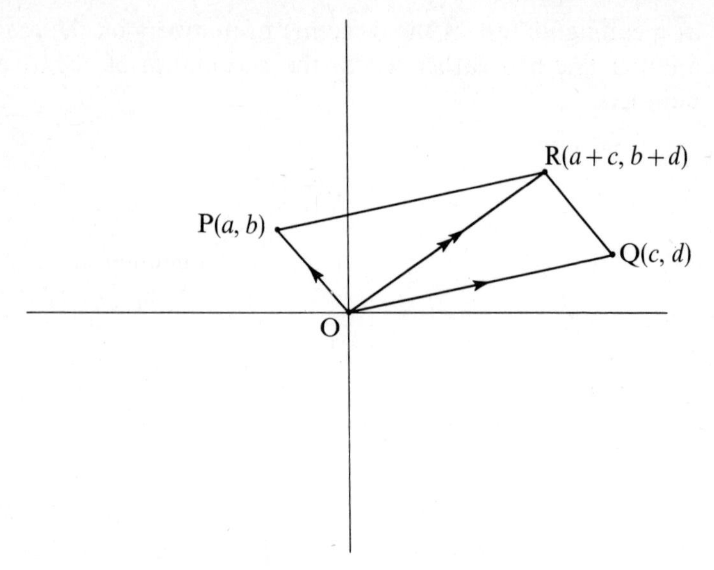

Addition of complex numbers

vector $\overrightarrow{OQ}$, in which case the addition formula

$$(a+bi)+(c+di) = (a+c)+(b+d)i$$

corresponds to the vector addition formula

$$\begin{aligned}\overrightarrow{OP}+\overrightarrow{OQ} &= \overrightarrow{OP}+\overrightarrow{PR}\\ &= \overrightarrow{OR}.\end{aligned}$$

In dealing with multiplication of complex numbers it seems best, as we have seen in a particular case, to think in terms of rotations. In considering the general geometrical interpretation to be given for multiplication we therefore use trigonometrical techniques, and polar co-ordinates (r, θ) rather than rectangular co-ordinates (x, y). Thus if $P(a, b)$, representing $a+bi$, is such that $\overrightarrow{OP}$ makes an angle θ with the positive real axis, and such that $\overrightarrow{OP}$ has length $r = \sqrt{(a^2+b^2)}$, so that P has polar co-ordinates (r, θ), we write

$$a+bi = r(\cos\theta+i\sin\theta),$$

and speak of $r(\cos\theta+i\sin\theta)$ as the 'polar form' of $a+bi$.

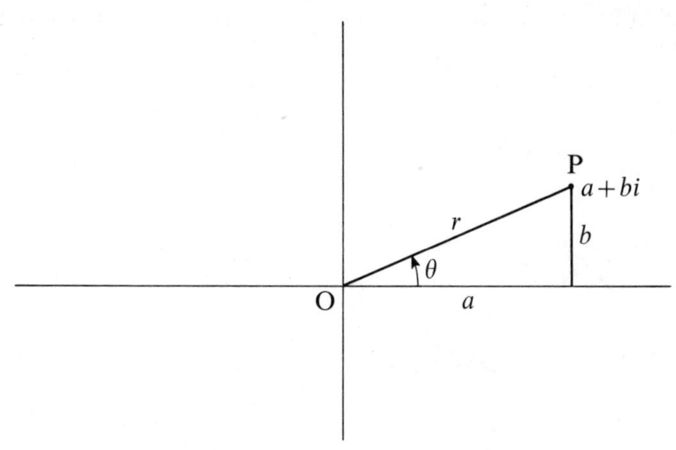

Polar form of complex numbers

If $s(\cos\phi+i\sin\phi)$ is another complex number in polar form, we use the addition formulae of elementary trignometry to obtain

$$(r(\cos\theta+i\sin\theta))(s(\cos\phi+i\sin\phi))$$

$$= rs(\cos\theta\cos\phi-\sin\theta\sin\phi)+i(\sin\theta\cos\phi+\cos\theta\sin\phi))$$

$$= rs(\cos(\theta+\phi)+i\sin(\theta+\phi)).$$

Here then we see in general the rotational effect of multiplication. When $r(\cos\theta+i\sin\theta)$ is multiplied by $s(\cos\phi+i\sin\phi)$, $\overrightarrow{OP}$ representing $r(\cos\theta+i\sin\theta)$ is rotated through an angle ϕ; at the same time a 'magnification' or 'dilation' occurs, and the length of the vector corresponding to the product of the two numbers is the product of the lengths of the vectors corresponding to the numbers: in polar co-ordinates we can write

$$(r,\theta)(s,\phi) = (rs,\theta+\phi)$$

It is perhaps interesting that despite these concrete geometrical interpretations of addition and multiplication of complex numbers we still perpetuate in our terminology the doubts our predecessors had about the 'reality' of complex numbers, by calling a the *real* part and b the *imaginary*

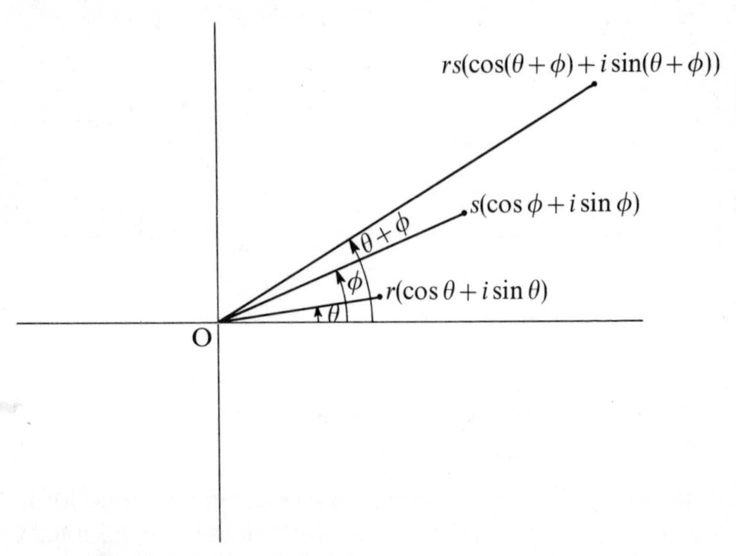

Multiplication of complex numbers in polar form

part of the complex number $a+bi$. In polar form, however, when

$$a+bi = r(\cos\theta+i\sin\theta),$$

we have alternative terminology; we call $r = \sqrt{(a^2+b^2)}$ the *absolute value*, or *modulus* of $a+bi$ and θ the *argument*, or *phase*. Some authors find it convenient to abbreviate $\cos\theta+i\sin\theta$ in the form $\text{cis}\,\theta$.

Putting geometrical interpretation on one side, when purely algebraic manipulation is required, we proceed automatically, using ordinary algebraic techniques, but replacing i^2 by -1 wherever it occurs. Thus for example we can simplify an expression such as $(1+i)^{-1}$ in the following way:

$$\frac{1}{1+i} = \frac{1-i}{(1+i)(1-i)}$$

$$= \frac{1-i}{(1-i^2)}$$

$$= \frac{1-i}{2}$$

$$= \tfrac{1}{2} - \tfrac{1}{2}i.$$

A little experience with such manipulations soon convinces one that the extension of one's traditional algebraic techniques to include the operation of replacing i^2 by -1 raises no difficulties.

Given then that we can manipulate complex numbers successfully and that we have real geometrical interpretations of the operations involved in these manipulations, what more is there to be said at this elementary level?

One could reasonably ask if this number system has any use, as it stands, outside mathematics. One answer might come from the electrical engineer who could draw attention to his use of elementary complex number theory in his analysis of circuits including inductances and capacitors. Others outside mathematics proper would join him; the modern scientist and engineer regard a knowledge of complex numbers as an essential part of their mathematical equipment.

On the other hand one might ask if there are any further basic pure mathematical questions which might be raised. Here, bearing in mind that since the middle of the last century mathematicians have increasingly recognised the need to study explicitly the abstract algebraic rules and properties of our various number systems, the modern mathematician would surely answer that there are questions concerned with the analysis of the algebraic rules governing the behaviour of the complex numbers which are worth asking. In modern parlance he would suggest that one should examine the *algebraic structure* of the complex numbers.

It is in this modern algebraic spirit that this book is written: in it we shall construct and analyse the complex numbers in a formal algebraic way, and in so doing we hope to exemplify elementary modern algebraic thinking.

We shall begin in the first chapter by considering in algebraic form the basic properties of the real numbers. This gives us a firm foundation for the algebraic theory to follow and will introduce the reader, in a reasonably familiar situation, to the kinds of basic algebraic properties which we recognise as important in studying number systems.

In our second chapter we observe that if an algebraic theory of the 'numbers' $a+bi$ is to be constructed at all, it must surely be constructed as a theory of pairs (a, b) of real numbers a and b. Thus, for example, the addition of complex numbers:

$$(a+bi)+(c+di) = (a+c)+(b+d)i,$$

can be written as an 'addition':

$$(a, b)+(c, d) = (a+c, b+d)$$

of such pairs of real numbers, without any reference to the 'imaginary' number i. We therefore begin our construction of the complex numbers by analysing the 'real' algebra of these pairs of real numbers when this rule of addition is imposed upon them. We are led in this analysis to a theory of the *two-dimensional real vector space* which constitutes an appropriate algebraic analogue of the geometrical theory of displacement vectors in the plane.

But if our analysis of this vector space of pairs of real numbers enables us to make explicit in algebraic terms the 'additive structure' of the complex numbers, and at the same time yields a geometrical interpretation on traditional lines, what can we say about multiplication? If we are to give our pairs of real numbers, or *vectors* as we now call them, a rule of multiplication which coincides with the complex rule:

$$(a+bi)(c+di) = (ac-bd)+(bc+ad)i,$$

we must insist on the rule:

$$(a, b)(c,d) = (ac-bd, bc+ad).$$

Is it obvious that this rule is a reasonable one to take in the context of our vector space? That it gives the multiplication we require is self-evident, but can we justify it as a 'reasonable' multiplication of vectors?

We answer these questions in our third chapter, where we list the algebraic criteria we require of a multiplication if it is to yield a satisfactory and 'reasonable' multiplicative structure in a vector space. We observe that our complex rule of multiplication satisfies these criteria, but even more than that, we find that this is the only rule we can take if our criteria are to be satisfied. In other words, we prove that the multiplication of the complex numbers is unique, in a strict algebraic sense, as a multiplication of two-dimensional real vectors.

Inevitably, we call our vector space, together with this multiplicative structure, the *complex numbers*, and note the 'real' way in which it has been constructed without reference to 'imaginary' numbers.

Finally, since our construction has been concerned with the algebraic equivalent of the 'cartesian co-ordinate' form (a, b) of $a+bi$, we examine our vector space again in Chapter 5 to see how we can introduce into it some algebraic equivalent of the traditional polar form $r(\cos\theta+i\sin\theta)$. Here we need two new ideas. First the theory of the 'length' of a vector (a, b), which corresponds to the traditional theory of the absolute value of a complex number, and secondly the theory of a 'rotation' of such a vector, which corresponds to the traditional theory of the argument of a complex number. The ideas involved here are geometrically obvious, and the appropriate algebraic theory in our vector space is naturally based on our geometrical understanding of length and angle. That one has to put in some fairly hard algebraic work to introduce the ideas into the vector space is a consequence of the fact that the appropriate algebraic structure is not built into our definition of the vector space of pairs. We need to 'enrich' the algebraic structure of the vector space before we can proceed to develop in it a theory of the polar form. Once we have done this, we have available algebraic techniques which enable us to interpret our multiplication of vectors, and hence of complex numbers, in rotational terms in our vector space.

We are thus led full circle by our construction and analysis: we find, in modern algebraic terms, that the additive and multiplicative structures of our complex numbers are concerned respectively with vector addition and rotation, whose geometrical interpretations are precisely those which led to the acceptance of the complex numbers. Apart from the benefit to be gained from a deeper understanding of the basic algebraic processes involved, we obtain en route a uniqueness theorem for the complex numbers. Such a theorem cannot be formulated, nor indeed could the question it answers be asked with the necessary precision, outside the conceptual framework of modern algebraic theory, which, as we have said, it is our aim to exemplify.

Naturally, throughout the book the formal treatment will be algebraic. Nevertheless at all times informal geometrical interpretations of the algebra will be made. The reader is therefore encouraged not only to follow the algebraic argument, but also to think geometrically and to draw illustrative pictures wherever possible. Such informal geometrical thinking should lighten the load of understanding the more formal algebra.

We close this introductory chapter with a few exercises in which the reader is invited to exercise his formal algebraic skill and geometrical understanding in manipulating 'numbers' of the form $a+bi$, where a and b are real and $i^2 = -1$.

Exercises

1 Express in the form $a+bi$, where a and b are real:

(i) $4(2+3i)$,

(ii) $(1+i)(1-i)$,

(iii) $(5+2i)^2$,

(iv) $(\cos\theta+i\sin\theta)(\cos\theta-i\sin\theta)$,

(v) $(x-\cos\theta-i\sin\theta)(x-\cos\theta+i\sin\theta)$, if x is real,

(vi) $\dfrac{1+3i}{1-2i}$

(vii) $\dfrac{(1-i)^2}{1+2i}$,
(viii) $(1+i)^{-1}+(1-2i)^3$,
(ix) $[\frac{1}{2}(-1\pm i\sqrt{3})]^3$.

2 Find the modulus and argument of
(i) $1+i$,
(ii) $\sqrt{3}-i$,
(iii) $i-1$,
(iv) $\cos\pi/4 - i\sin\pi/4$.

3 If a given point P in the plane corresponds to a complex number $z = a+bi$, construct the points corresponding to
(i) $-2z$, (ii) $z+5$, (iii) $7-3z$, (iv) z^2, (v) z^{-1}.

CHAPTER ONE

The real number field $\mathbf{R}$

1.1 An algebraic view of the integers

When we first begin to study algebra we replace the 'numbers' which we have learnt to use in arithmetic by 'letters', and then, by extrapolating from our arithmetical experience, we collect together appropriate sets of rules which the letters will have to obey if they are to behave in the same way as numbers. In attempting to recognise again the formal rules of algebra, we therefore cast our minds back to the very simplest of arithmetical facts which we associate with our various kinds of numbers.

Consider from this point of view the integers

$$\ldots, -2, -1, 0, 1, 2, \ldots$$

We know from experience that we can *add* and *multiply* integers with gay abandon, obtaining as a result integer answers. Moreover in adding a collection of integers we can rearrange them in any order, and we can do the individual additions which make up the calculation in any order. The same is also true in multiplication. Thus, for example, by rearrangement:

$$3+2+7+(-5) = (-5)+2+7+3,$$

and $$3\times 2\times 7\times(-5) = (-5)\times 2\times 7\times 3;$$

or again, using different bracketing in the usual way to indicate how to perform the individual parts of the 'sums':

$$(3+2)+(7+(-5)) = 3+(2+(7+(-5))),$$

and $$(3\times 2)\times(7\times(-5)) = 3\times(2\times(7\times(-5))).$$

We can also reverse the process of addition, and subtract integers, again obtaining integer answers. Moreover, this reversal of addition is always possible: not only can we subtract 2 from 4, but also 4 from 2, since we have available not only the positive, but also the negative integers.

In multiplication the situation is different. Certainly we can always multiply integers by integers and obtain integer answers; probably the only care we need to take is in recognising the various rules of signs, such as 'a minus times a minus is a plus'. But we must also recall phrases such as '4 into 2 won't go', so that although the reverse of multiplication, namely division, is sometimes possible (e.g. 2 does 'go into' 4, twice), we should not expect it always to be so. Thus we may need to take more care when discussing 'division' and 'inverses' of integers than when we discuss the corresponding additive concepts of 'subtraction' and 'negation'.

We can translate these arithmetical facts into formal algebraic language as follows. First, for addition, we must assert a 'rule' that given any two integers m, n there is defined a third integer called their *sum* and written $m+n$. This we can paraphrase if we wish by saying *the integers are closed under addition*, by which we mean that adding two integers produces another integer. We must also note that $m+n = n+m$, so that we say *the process of addition of integers is commutative*, meaning we can add integers in any order and not affect the result. Finally we must note that

$$(m+n)+r = m+(n+r);$$

here we say *addition of integers is associative*, meaning that the order in which one performs the process of addition does not affect the result, so that we write the sum of the integers m, n, r in the form $m+n+r$, omitting the brackets.

To add more than two integers we repeat the basic process. Thus, as we have seen, because of associativity we can define

$$m_1+m_2+m_3 = (m_1+m_2)+m_3,$$

and we can then proceed to define

$$m_1+m_2+m_3+m_4 = (m_1+m_2+m_3)+m_4$$

and then

$$m_1+m_2+m_3+m_4+m_5 = (m_1+m_2+m_3+m_4)+m_5,$$

or in general (for $k \geqslant 1$),

$$m_1+m_2+\ldots+m_{k+1} = (m_1+m_2+\ldots+m_k)+m_{k+1}.$$

Of course, having made this general definition, we must then note the generalized forms of the commutativity and associativity conditions: for commutativity, the sum of any number of integers is independent of their order; for associativity, the sum is independent of the way the integers may be grouped together in brackets.

Now consider subtraction. Here we begin our algebraic formalization by defining the 'negative' of an integer, and this first requires that we recognize the special integer *zero*. We therefore assert that there is one and only one integer, written 0, with the property that $0+m = m+0 = m$ for all integers m. We say 0 is the *additive identity*, meaning that it leaves any integer unaltered when added to it. The *negative* of any integer m, written $-m$, is then the unique integer with the property that $(-m)+m = m+(-m) = 0$. One can refer to $-m$ as the *additive inverse* of m, meaning that when it is added to m the resulting sum is the additive identity 0. Finally we define subtraction in terms of our basic operation of addition: we subtract n from m by adding $-n$ to m, and the result, $m+(-n)$, is of course written $m-n$.

These ideas can also be dealt with geometrically using the standard pictorial illustration of addition and subtraction in terms of appropriate *displacements* on a number line marked out in the usual way by *places* ..., '-2', '-1', '0', '1', '2', ...

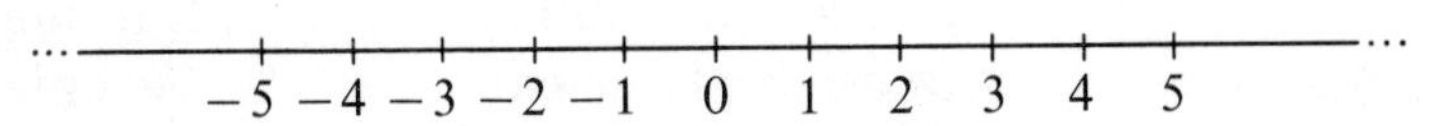

By a displacement m we mean a specification to move m places, to the right if m is one of the integers 1, 2, 3, ..., to the left if it is one of -1, -2, -3, ..., and not to move at all if $m = 0$. The actual *movement* which we use to 'represent' a given displacement will depend on the starting place we choose. Thus given a displacement 3, we could start at place '2' and move to place '5', or we could start at place '-6' and move to place '-3'; there are an infinity of possibilities. We take care therefore to distinguish between our use of the word *displacement* and the word *movement*. Summing up, *movements along the number line represent displacements: movements have specified starting points, displacements do not.*

There is an obvious 'addition' of displacements corresponding precisely to our addition of integers. For example, if we have to move 3 places and then 2 more, we have to move 5 places in all. Instead of calling this an 'addition' of displacements, we usually speak of a *composition* of displacements; in the example, the displacement 3 is composed with the displacement 2 to obtain the displacement 5. Clearly, in associating movements with the displacements in this situation, it is most useful if we arrange that the movement representing the displacement 2 starts at the finishing place of the movement representing the displacement 3. Only then can the *movements* be composed as we require if we are to obtain the most suitable and useful geometrical illustration.

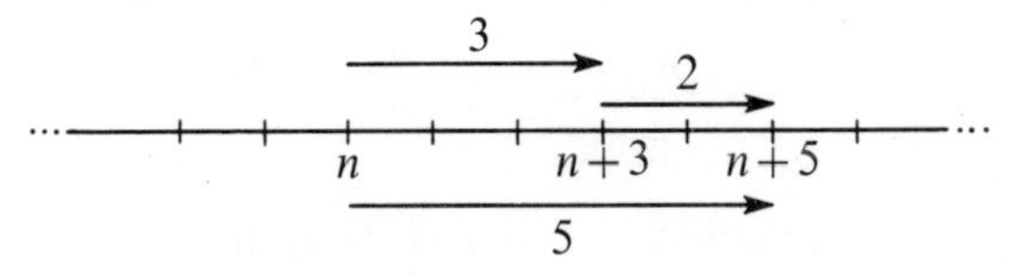

($5=3+2$: illustrated by movements from place 'n')

Not only our rule for addition, but also all our other rules can now be suitably interpreted. Thus the uniqueness of the integer 0 in relation to addition corresponds to the uniqueness of the zero displacement, which is the only displacement represented by no movement, so that when it is composed with any other displacement it does not change that displacement. Again, the negation of integers corresponds to the reversal of the direction of displacements. The algebraic relationship between negation and zero, namely $m+(-m) = 0$, is carried over as it should be, for if we compose the displacement m with the displacement $-m$, we move m places in one direction and then m in the opposite direction, and we have in the end not displaced ourselves at all, i.e. we have obtained the zero displacement.

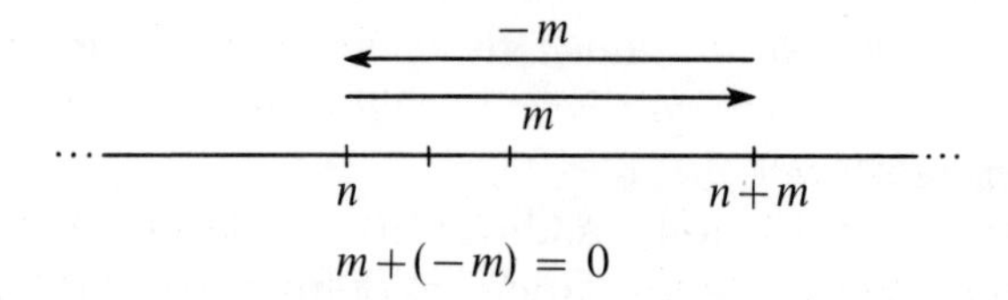

$m+(-m) = 0$

We now leave what we call the rules concerning the *additive structure* of the integers and turn to the algebraic formalization of the process of multiplication. Again we can begin with a *closure* rule: *the integers are closed under the operation of multiplication*, i.e. given any two integers m, n, their product is again an integer, written mn. We must then note the *commutativity of multiplication*, namely $mn = nm$, and *associativity*, namely $m(nr) = (mn)r$, for all integers m, n, r.

To multiply together more than two integers, once again we repeat the basic process. First, bearing in mind associativity of multiplication, we define

$$m_1m_2m_3 = (m_1m_2)m_3,$$

and then proceed to define

$$m_1m_2m_3m_4 = (m_1m_2m_3)m_4,$$

$$m_1m_2m_3m_4m_5 = (m_1m_2m_3m_4)m_5,$$

and so on, so that in general for $k \geqslant 1$ we set

$$m_1 m_2 \ldots m_k m_{k+1} = (m_1 m_2 \ldots m_k) m_{k+1}.$$

Having made this general definition we must also again note the generalized forms of commutativity and associativity: the product of any number of integers is independent on the one hand of their order, and, on the other, of the way they may be grouped in brackets.

Before going any further with the multiplication of integers, let us observe how the two operations of addition and multiplication are inter-related by what we call the rules of *distribution*: for all integers m, n and r,

$$m(n+r) = mn+mr,$$

and $$(m+n)r = mr+nr.$$

Of course there are again generalized forms for these rules, which can be stated in one general distributive rule:

$$(m_1+m_2+\ldots+m_k)(n_1+n_2+\ldots+n_l) = m_1n_1+\ldots+m_1n_l+\ldots+m_kn_1+\ldots+m_kn_l.$$

Returning to our previous line of argument concerned with multiplication alone, we must next consider what we can say about the reverse of multiplication of integers. Although, as we have already said, we cannot hope for a theory as complete as that given in our algebraic treatment of subtraction considered as the reverse of addition, nevertheless we can start on lines similar to those we followed in dealing with subtraction. To begin with, then, we need an integer which behaves with respect to multiplication as 0 behaves with respect to addition, i.e. leaves any integer unaltered when it is multiplied by it. Clearly 1 is the unique integer such that $1m = m1 = m$ for all integers m, so just as we have an additive identity 0, we have a *multiplicative identity* 1. Writing the product mn as $m \times n$, the similarity of the two situations becomes obvious, since we then have

$$m+0 = 0+m = m$$

and $$m \times 1 = 1 \times m = m.$$

So far, so good! Again, therefore, following lines similar to those we followed in dealing with subtraction, we next reconize that we need 'multiplicative inverses'. Recall that additive inverses can be found for all integers, i.e. given any integer m there exists an integer x, namely $x = -m$, such that

$$m+x = x+m = 0.$$

This same statement written multiplicatively should read 'given any integer m there exists an integer x such that

$$m \times x = x \times m = 1.\text{'}$$

But only if $m = 1$ or -1 can we find an integer x satisfying this equation; for general m there is no such integer x.

What then can we do about reversing the technique of multiplication of integers? Clearly we cannot set up a theory following the pattern of argument we used in dealing with subtraction as the reverse of addition. The structural similarity between addition and multiplication ends at this point when we are studying the algebra of the integers. Only if we are allowed to 'extend' our number system by regarding the integers as a part of the larger system of rational numbers, which give fractional solutions $x = 1/m$ to such equations as $mx = 1$, can we hope to preserve a similarity of additive and multiplicative structure.

Later in this chapter we shall carry out this extension and study the rational numbers from an algebraic point of view. Then we shall be able to invert integers, associating the rational number $1/m$, or m^{-1} as we usually write it, with any non-zero integer m. In the case of the rational numbers we shall then have strong similarities between addition and multiplication: 0 compares with 1; negation, $-m$, with inversion, m^{-1}; subtraction, $m-n$, with division, m/n, etc.

But before going to these more general lengths, we can develop appropriate algebraic techniques for dealing with the reversal of integer multiplication, *when this is possible.* First then, let us attempt to recognize when we do and when we do not reverse the procedures of integer multiplication. At the heart of our problem lies the arithmetic difference between equations such as $3x = 1$ and $3x = 12$. In the former, as we might say at an elementary level, '3 does not go into 1', in the latter, since $4 \times 3 = 12$, we say '3 does go into 12', and in saying 'it goes 4 times' we reverse the process of multiplication; in other words, $4 \times 3 = 12$ implies $12 \div 4 = 3$. We can solve $3x = 12$, obtaining the integer answer $x = 4$. We cannot so solve $3x = 1$.

In practice how do we solve the equation $3x = 12$? Surely by cancelling 3 on both sides of the equation to obtain $x = 4$. Additively we have a similar process, when we say, for example, $3+x = 12 = 3+9$, whence 'cancelling 3', we have the solution $x = 9$. However, once again this last process of additive cancellation is always possible, as is shown by the following justification (which, it should be noted, is based entirely on the additive rules for integers which we have previously made explicit).

If $\qquad m+a = m+b,$

then $\qquad (-m)+(m+a) = (-m)+(m+b)$

(add $(-m)$ to both sides)

whence $\qquad ((-m)+m)+a = ((-m)+m)+b$

(by associativity)

i.e. $\qquad 0+a = 0+b$

(since $(-m)+m = 0$)

i.e. $\qquad a = b$

(since $0+a = a$, and $0+b = b$).

Now try writing this argument multiplicatively, i.e. replacing addition by multiplication, $(-m)$ by m^{-1} and 0 by 1.

If $\qquad ma = mb,$

then $\qquad (m^{-1})(ma) = (m^{-1})(mb)$

(multiply both sides by m^{-1})

whence $(m^{-1}m)a = (m^{-1}m)b$

(by associativity)

i.e. $1a = 1b$

(since $m^{-1}m = 1$)

i.e. $a = b$

(since $1a = a$, and $1b = b$).

This multiplicative argument is of course, as we expect, not allowable within the integers, for we cannot assume that for all integers m there is an *integer* m^{-1}. Nevertheless, when $m \neq 0$, we know from our arithmetical experience that '$ma = mb$' does imply '$a = b$' and we need an algebraic rule to justify this multiplicative cancellation of integers. The one we take arises from the observation that given any two integers m and n such that $mn = 0$, either $m = 0$ or $n = 0$, or both, is true. With this observation as a new rule, we have a correct justification for 'multiplicative cancellation.':

$$ma = mb$$

implies $ma - mb = 0,$

i.e. $m(a-b) = 0,$

whence $m = 0$, or $a - b = 0$. Thus if $m \neq 0$, we have $a - b = 0$, or $a = b$, so that '$ma = mb$, $m \neq 0$' implies '$a = b$', as required for multiplicative cancellation.

Of course, as we have implied earlier, the use of inverses in multiplicative cancellation, which we have seen to be improper for integers, becomes possible when we extend our number system and work with the rational and the real numbers. We shall consider the rules and properties of these two larger number systems in the next two sections of this chapter, and in so doing we shall observe in detail the various advantages we gain when we no longer restrict our attention to the integers. In the meantime we now summarize the more important of the fact we have collected concerning the behaviour of the integers.

(i) Given any integers m, n their sum $m+n$ and product mn are uniquely defined integers;

(ii) for all integers m, n,

$$m+n = n+m, \quad mn = nm;$$

(iii) for all integers m, n, r,

$$m+(n+r) = (m+n)+r, \quad m(nr) = (mn)r;$$

(iv) for all integers m, n, r,

$$m(n+r) = mn+mr, \quad (m+n)r = mr+nr;$$

(v) there is a unique integer 0 such that for all integers m,

$$0+m = m = m+0;$$

(vi) there is a unique integer 1 such that for all integers m,

$$1m = m = m1;$$

(vii) given any integer m there exists a unique integer $-m$ such that

$$m+(-m) = 0 = (-m)+m;$$

(viii) if m, n, r are integers such that $m \neq 0$ and $mn = mr$, then $n = r$.

Worked examples 1.1

1 Even integers are combined by a rule '*' defined by

$$m * n = 2mn.$$

Are the even integers closed under this rule of combination? Is the rule commutative, associative and distributive over ordinary addition? Is there an 'identity' for this rule of combination?

Since $2mn$ is even whatever integers m and n are taken, certainly the even integers combine under * to give another even integer, and are therefore closed under the rule '*'.

Since $2mn = 2nm$, we have $m * n = n * m$, and '$*$' is commutative.

The rule is also associative, since

$$\begin{aligned}(m * n) * p &= (2mn) * p \\ &= 2((2mn)p) \\ &= 4mnp \\ &= 2(m(2np)) \\ &= m * 2np \\ &= m * (n * p).\end{aligned}$$

Distributivity over ordinary addition also follows since

$$\begin{aligned}m * (n+p) &= 2m(n+p) \\ &= 2mn+2mp \\ &= (m * n)+(m * p),\end{aligned}$$

and

$$\begin{aligned}(m+n) * p &= 2(m+n)p \\ &= 2mp+2np \\ &= (m * p)+(n * p).\end{aligned}$$

Finally, we require an even integer e with the property $e * m = m * e = m$ for all even integers m. But if $e * m = m$, we have $2em = m$, and hence, taking $m \neq 0$, $2e = 1$, which is not possible since the integer 2 does not have an integer inverse. Thus the rule '$*$' has no identity.

2 Integers are combined by the rule '$*$', defined by

$$m * n = m - n.$$

Is this rule associative?

No, since taking $m = n = 0$ and $p = 1$, we have

$$
\begin{aligned}
m * (n * p) &= 0 * (0-1) \\
&= 0-(0-1) \\
&= 1,
\end{aligned}
$$

whereas
$$
\begin{aligned}
(m * n) * p &= (0-0)-1 \\
&= -1,
\end{aligned}
$$

so that for suitable choice of m, n and p we have

$$m * (n * p) \neq (m * n) * p.$$

3[†] Prove '$m0 = 0m = 0$, for any integer m', using only the following 'rules' for integers:

(i) $m+0 = m = 0+m$,
(ii) $m(p+q) = mp+mq$,
(iii) $(m+p)q = mq+pq$,
(iv) $m+(p+q) = (m+p)+q$,
(v) given any integer m, there exists a unique integer, $(-m)$ such that

$$(-m)+m = 0 = m+(-m).$$

Since $m(m+0) = mm$ (by (i))

we have $mm+m0 = mm$ (by (ii)).

Hence $(-mm)+(mm+m0) = (-mm)+mm$,

i.e. $((-mm)+mm)+m0 = (-mm)+mm$ (by (iv))

and $0+m0 = 0$, (by (v))

whence $m0 = 0$ (by (i)).

Similarly

$$(0+m)m = mm,$$

so that $0m+mm = mm$,

[†] The reader should note that in questions such as this, where 'rules' are specified, a strict logical argument is required.

and 'cancelling mm', we have

$$0m = 0.$$

4[†] Prove '$m(-n) = (-m)n = -(mn)$ for any integers m, n', using only the result of example 3 above, and 'rules' (ii), (iii) and (v) of that example.

We require to show that $m(-n)$ and $(-m)n$ are both solutions of the equation

$$mn + x = 0,$$

since rule (v) then implies that each of $m(-n)$ and $(-m)n$ is $-(mn)$.

We have

$$
\begin{aligned}
mn + m(-n) &= m(n + (-n)) && \text{(Rule (ii))} \\
&= m0 && \text{(Rule (v))} \\
&= 0 && \text{(Example 3 above).}
\end{aligned}
$$

Similarly using (iii) in place of (ii), we can prove $(-m)n = -(mn)$.

Exercises 1.1

1 In each of the following cases, a rule '$*$' is given for combining integers. Discuss the closure of the integers, the commutativity, associativity, distributivity over addition, and the existence of an identity, and inverses, for the rule $*$.

(i) $m * n = \frac{1}{2}(m+n)$,
(ii) $m * n = m + 2n$,
(iii) $m * n = 1 + mn$,
(iv) $m * n = -mn$,
(v) $m * n = m^n$.

† The footnote referring to Example 3 also applies here.

2 Deduce† from the following 'rules' for integers:
(i) $a(b+c) = ab+ac$,
(ii) $(a+b)c = ac+bc$,
(iii) $a(-b) = (-a)b = -(ab)$,
that

$$a(b-c) = ab-ac,$$

and

$$(a-b)c = ac-bc,$$

for any integers a, b, c.

1.2 An algebraic view of the rational numbers

We now consider rational numbers from an algebraic point of view, aiming in particular, given integers m, n, to make algebraic sense of the 'numbers' mn^{-1}, which we have seen in Section 1.1 cannot always be integers. As in the case of the integers, we shall deal first with the rules for additive structure and then with the rules for multiplicative structure, at each stage picking out those algebraic rules of behaviour which appear to be basic in our understanding of how these numbers behave.

Let us begin by recalling that we write rational numbers in fractional form, m/n, where m and n are integers and n is not zero. However, we want to agree that equations such as the following are true:

$$\frac{1}{2} = \frac{2}{4} = \frac{-4}{-8} = \frac{17}{34}, \text{ etc.}$$

What these equations mean is that all these fractions represent the same rational number, namely one-half. The explicit and general algebraic result is that *fractions m/n and p/q represent the same rational number if and only if the integers mq and np are equal.*

Addition of rational numbers is then written in terms of addition and multiplication of the integers in the 'numerators' and 'denominators' of the fractions. To see what the

† Strict deduction again required here!

rule should be, we note that it would seem to be reasonable to define in the following special case:

$$\frac{a}{n}+\frac{c}{n} = \frac{a+c}{n}$$

In the general case we wish to define $\frac{a}{b}+\frac{c}{d}$. However,

$$\frac{a}{b}+\frac{c}{d} = \frac{ad}{bd}+\frac{bc}{bd}.$$

This leads to our definition

$$\frac{a}{b}+\frac{c}{d} = \frac{ad+bc}{bd}.$$

To distinguish for the time being between rational and integer addition we shall use the symbol $\oplus$ for the former and the usual symbol $+$ for the latter. We therefore write our definition of rational addition in the form

$$\frac{a}{b} \oplus \frac{c}{d} = \frac{ad+bc}{bd}.$$

Question 1. The rule for addition of two rational numbers is written using fractional representations for the rational numbers. Is the sum dependent on these representations?

Answer. No; for example, replace $\frac{a}{b}$ by $\frac{a'}{b'}$ where $\frac{a}{b}$ and $\frac{a'}{b'}$ represent the same rational number, i.e. are such that the integers ab' and ba' are equal. We then have

$$\frac{a'}{b'} \oplus \frac{c}{d} = \frac{a'd+b'c}{b'd}.$$

But
$$\frac{a'd+b'c}{b'd} = \frac{ad+bc}{bd},$$

since
$$(a'd+b'c)(bd) = (a'd)(bd)+(b'c)(bd),$$

(by distributivity of integers)

$= a'dbd + b'cbd$

(by associativity of multiplication of integers)

$= a'bdd + b'dbc$

(by commutativity of multiplication of integers)

$= ab'dd + b'dbc$

(since $a'b = ab'$)

$= b'dad + b'dbc$

(by commutativity of multiplication of integers)

$= b'd(ad + bc)$

(by distributivity of integers).

In short, replacing $\frac{a}{b}$ by an 'equal' fraction $\frac{a'}{b'}$ replaces $\frac{ad+bc}{bd}$ by an 'equal' fraction $\frac{a'd+b'c}{b'd}$. The same holds true if we replace $\frac{c}{d}$ by $\frac{c'}{d'}$.

We have thus deduced the all-important fact that any suitable representative fractions can be used in formulae relating to addition of rationals. It is because of the importance of this fact that we have taken such care to base each step in our deduction on the facts we know about the algebraic behaviour of the integers; the detailed proof, spelling out all the appropriate algebraic manipulation without 'short cuts', is inevitable if we are to identify the use of our rules for integers at each stage.

We next turn to the general rules governing our addition.

Question 2. Is addition of rational numbers commutative, like addition of integers?

Answer. Yes, since

$$\frac{a}{b} \oplus \frac{c}{d} = \frac{ad+bc}{bd}$$

$$= \frac{da+cb}{db}$$

(by commutativity of multiplication of integers)

$$= \frac{cb+da}{db}$$

(by commutativity of addition of integers)

$$= \frac{c}{d} \oplus \frac{a}{b}$$

(by the rule for rational addition).

Question 3. Is addition of rational numbers associative?

Answer. Yes, at least in the simplest case, since

$$\left(\frac{a}{b} \oplus \frac{c}{d}\right) \oplus \frac{e}{f} = \frac{ad+bc}{bd} \oplus \frac{e}{f}$$

$$= \frac{(ad+bc)f+(bd)e}{(bd)f}$$

(by the rule for rational addition)

$$= \frac{((ad)f+(bc)f)+(bd)e}{(bd)f}$$

(by distributivity of integers)

$$= \frac{(a(df)+b(cf))+b(de)}{b(df)}$$

(by associativity of multiplication of integers)

$$= \frac{a(df)+(b(cf)+b(de))}{b(df)}$$

(by associativity of addition of integers)

$$= \frac{a(df)+b(cf+de)}{b(df)}$$

(by distributivity of integers)

$$= \frac{a}{b} \oplus \frac{cf+de}{df}$$

(by the rule for rational addition)

$$= \frac{a}{b} \oplus \left(\frac{c}{d} \oplus \frac{e}{f}\right)$$

(by the rule for rational addition).

Question 4. Is there an additive identity, or 'zero' for addition of rational numbers?

Answer. Yes, $\frac{0}{1}$ is such an identity,[†] since

$$\frac{a}{b} \oplus \frac{0}{1} = \frac{a1+b0}{b1}$$

(by the rule for rational addition)

$$= \frac{a+0}{b}$$

(since for integers, $a1 = a$, $b0 = 0$)

$$= \frac{a}{b}$$

(since for integers $a+0 = a$),

[†] Here we need to recall that in our fractional form m/n, the integer n is never allowed to be 0. We thus ought to note that if in collecting facts about integers we had hoped to be comprehensive we should have stated the 'rule' $1 \neq 0$, for use at this point.

and, by commutativity, it follows that $\frac{0}{1} \oplus \frac{a}{b} = \frac{a}{b}$, as also required.

Question 5. Is this additive identity, $\frac{0}{1}$, unique, in the sense that if $\frac{a}{b} \oplus \frac{x}{y} = \frac{a}{b}$ for all $\frac{a}{b}$, then $\frac{x}{y} = \frac{0}{1}$?

Answer. Yes, since if $\frac{a}{b} \oplus \frac{x}{y} = \frac{a}{b}$ for all $\frac{a}{b}$, we can take $\frac{a}{b} = \frac{0}{1}$, and obtain

$$\frac{0}{1} \oplus \frac{x}{y} = \frac{0}{1}.$$

But

$$\frac{0}{1} \oplus \frac{x}{y} = \frac{x}{y},$$

(from answer to Question 4)

so that

$$\frac{x}{y} = \frac{0}{1},$$

as required.

Question 6. Is there an additive inverse, or 'negative' for any rational number?

Answer. Yes, given $\frac{a}{b}$, we can take $\frac{-a}{b}$ as an additive inverse, since

$$\frac{a}{b} \oplus \frac{-a}{b} = \frac{ab+b(-a)}{bb}$$

(by rule for rational addition)

$$= \frac{ba+b(-a)}{bb}$$

(by commutativity of multiplication of integers)

$$= \frac{b(a+(-a))}{bb}$$

(by distributivity of integers)

$$= \frac{b0}{bb}$$

(since for integers $a+(-a) = 0$)

$$= \frac{0}{bb}$$

(since for integers $b0 = 0$)

$$= \frac{0}{1}$$

(since for integers $01 = (bb)0 = 0$),

and, by commutativity, $\frac{-a}{b} \oplus \frac{a}{b} = \frac{0}{1}$ also.

Question 7. Is this additive inverse, $\frac{-a}{b}$, unique, in the sense that if $\frac{a}{b} \oplus \frac{x}{y} = \frac{0}{1}$, then $\frac{x}{y} = \frac{-a}{b}$?

Answer. Yes, for suppose

$$\frac{a}{b} \oplus \frac{x}{y} = \frac{0}{1},$$

then $$\frac{-a}{b} \oplus \left(\frac{a}{b} \oplus \frac{x}{y}\right) = \frac{-a}{b} \oplus \frac{0}{1},$$

whence $$\left(\frac{-a}{b} \oplus \frac{a}{b}\right) \oplus \frac{x}{y} = \frac{-a}{b}$$

(by associativity and property of $\frac{0}{1}$)

But as we have noted in answer to Question 5 above,

$\dfrac{-a}{b} \oplus \dfrac{a}{b} = \dfrac{0}{1}$, so that

$$\begin{aligned}\frac{x}{y} &= \frac{0}{1} \oplus \frac{x}{y}\\ &= \left(\frac{-a}{b} \oplus \frac{a}{b}\right) \oplus \frac{x}{y}\\ &= \frac{-a}{b},\end{aligned}$$

as required. Following the usual practice we now write this additive inverse $\dfrac{-a}{b}$ in the form $-\dfrac{a}{b}$.

We are thus led to the conclusion that the rational numbers behave additively in exactly the same way as the integers.† What can we now say about their multiplication? Our usual arithmetic rule for multiplication of rationals, expressed in fractional algebraic terms, takes the form

$$\left(\frac{m}{n}\right)\left(\frac{p}{q}\right) = \frac{mp}{nq}.$$

Again we must check that this rule is independent of the fractional representations. If therefore we replace $\dfrac{m}{n}$ by $\dfrac{m'}{n'}$ where $mn' = nm'$, we note

$$\left(\frac{m'}{n'}\right)\left(\frac{p}{q}\right) = \frac{m'p}{n'q}.$$

But $(m'p)(nq) = (n'q)(mp)$, by commutativity and associativity of multiplication of integers. Thus

$$\frac{m'p}{n'q} = \frac{mp}{nq},$$

† We take for granted (but see Appendix 1) that associativity and commutativity in their simplest forms imply commutativity and associativity in general when more than two or three rational numbers are added together.

as required. Similarly, if we replace $\frac{p}{q}$ by $\frac{p'}{q'}$, we have

$$\frac{mp'}{nq'} = \frac{mp}{nq}.$$

We shall leave as exercises (to be set later) the proofs that
(i) multiplication of rational numbers is commutative† meaning that

$$\left(\frac{m}{n}\right)\left(\frac{p}{q}\right) = \left(\frac{p}{q}\right)\left(\frac{m}{n}\right);$$

(ii) multiplication of rational numbers is associative,† meaning that

$$\left(\frac{m}{n}\right)\left(\left(\frac{p}{q}\right)\left(\frac{r}{s}\right)\right) = \left(\left(\frac{m}{n}\right)\left(\frac{p}{q}\right)\right)\left(\frac{r}{s}\right);$$

(iii) distributivity holds for rational numbers, meaning that

$$\left(\frac{m}{n}\right)\left(\frac{a}{b} \oplus \frac{c}{d}\right) = \left(\frac{m}{n}\right)\left(\frac{a}{b}\right) \oplus \left(\frac{m}{n}\right)\left(\frac{c}{d}\right)$$

and
$$\left(\frac{m}{n} \oplus \frac{p}{q}\right)\left(\frac{a}{b}\right) = \left(\frac{m}{n}\right)\left(\frac{a}{b}\right) \oplus \left(\frac{p}{q}\right)\left(\frac{a}{b}\right).$$

That there is a multiplicative identity, namely $\frac{1}{1}$, for rational numbers is trivial, for

$$\left(\frac{m}{n}\right)\left(\frac{1}{1}\right) = \frac{m1}{n1} = \frac{m}{n},$$

since for integers m, n, $m1 = m$ and $n1 = n$.

That this identity is unique is also easily proved by translating into multiplicative terms the additive argument on page 33. Thus if, for all $\frac{m}{n}$, we have

$$\left(\frac{m}{n}\right)\left(\frac{x}{y}\right) = \frac{m}{n},$$

† Again, see Appendix 1. We assume for the present that these 'simplest cases' of commutativity and associativity imply the general ones, once addition and multiplication have been defined in general.

it follows that we may take $m = n = 1$, whence

$$\left(\frac{1}{1}\right)\left(\frac{x}{y}\right) = \frac{1}{1},$$

i.e.
$$\frac{1x}{1y} = \frac{1}{1}.$$

But by our rules for the multiplication of integers $1x = x1 = x$, and $1y = y1 = y$, so that $\frac{x}{y} = \frac{1}{1}$, as required.

Finally we can take the step beyond the point we reached in considering the integers. We can observe that if $\frac{m}{n} \neq \frac{0}{1}$, i.e. if $m = m1 \neq n0$, then we have, associated with the fraction $\frac{m}{n}$, the fraction $\frac{n}{m}$, representing a rational number since m is not zero, and such that

$$\left(\frac{m}{n}\right)\left(\frac{n}{m}\right) = \frac{mn}{nm}$$

$$= \frac{1}{1} \text{ (since } mn = nm).$$

In other words, *non-zero rational numbers have multiplicative inverses*. Naturally we write the multiplicative inverse of $\frac{m}{n}$ as $\left(\frac{m}{n}\right)^{-1}$ and note that the process of multiplicative inversion is again a unique process, in that if

$$\left(\frac{m}{n}\right)\left(\frac{x}{y}\right) = \frac{1}{1},$$

then
$$\frac{mx}{ny} = \frac{1}{1},$$

i.e. $(mx)1 = (ny)1,$

so $mx = ny$

(by properties of 1)

whence $$xm = yn$$

(by multiplicative commutativity of integers)

so that $$\frac{x}{y} = \frac{n}{m}.$$

It is most important to notice that the rational numbers which can be written in the fractional form $\frac{m}{1}, \frac{n}{1}$, etc., where m, n, etc., are any integers, behave in a manner completely indistinguishable from that of integers. Thus

$$\frac{m}{1} \oplus \frac{n}{1} = \frac{m+n}{1}$$

and $$\left(\frac{m}{1}\right)\left(\frac{n}{1}\right) = \frac{mn}{1},$$

so that apart from notation ($\frac{m}{1}$ for example, rather than m) we shall not be able to distinguish between such rational numbers and the integers. We therefore write $m, n, \ldots$, for $\frac{m}{1}, \frac{n}{1}, \ldots$, and regard the rational numbers as a number system which contains the integers, or equivalently we say that the integers are embedded in the rationals. We now drop the notation $\oplus$ for addition of rationals and use $+$ throughout for both integer and rational addition. We also write mn^{-1} for $\frac{m}{n}$ whenever convenient. Notice that we should not, for complete accuracy, talk about 'the rational number $\frac{m}{n}$', but instead say 'the rational number with fractional representative $\frac{m}{n}$'. However the last phrase is too long to bear repetition and so we agree to let $\frac{m}{n}$ denote also the rational number represented by $\frac{m}{n}$. In particular, $\frac{1}{2}$, $\frac{2}{4}$, 2^{-1}, or in decimal notation 0·5, will all denote the rational number one-half.

In the larger, rational, system we can find inverses for all non-zero numbers, so we can say that the non-zero rational numbers are *closed* under multiplicative inversion, whereas the non-zero integers are not. More loosely, we sometimes say that the rationals have a 'richer' algebraic structure than the integers. In particular because of this 'richer structure' we can now employ the multiplicative inverse 'r^{-1}', to deduce $a = b$ from the equation $ra = rb$ in case r is any non-zero rational number; moreover our argument will even work if r is an 'integer' $m = \frac{m}{1}$. Thus we multiply both sides of the equation

$$ra = rb$$

by r^{-1} (which exists because $r \neq 0$) and then proceed as before to use associativity:

$$r^{-1}(ra) = r^{-1}(rb)$$

whence $(r^{-1}r)a = (r^{-1}r)b$ (by associativity)

so that $1a = 1b.$

But $1a = a$ and $1a = b,$

whence $a = b.$

We recall, that as we observed on page 23, such an argument cannot be used *within the integers*, for there are in general no integer inverses m^{-1} for integers $m \neq 0$. However, the enrichment and enlargement of the integers which takes place when we move to the rational numbers enables us to use this argument even when the rational number r is an 'integer' $\frac{m}{1}$, $m \neq 0$.

This process, replacing one number system (in this case the integers) by a larger one (in this case the rationals), making previously false and inapplicable arguments possible, is one of general application. Notice particularly that we have *constructed* the larger system out of the smaller one: the rationals have been defined by means of their fractional representations $\frac{m}{n}$, which are simply pairs of integers given in the order m, n, and moreover addition and multiplication of rationals have been defined and justified in terms of addition and multiplication of integers.

Our construction of rationals from pairs of integers has a geometrical counterpart: given the number line marked off with the integer points, ..., -2, -1, 0, 1, 2, ..., we can give a geometrical, compass and ruler, construction for finding an appropriate point to represent any rational $\frac{m}{n}$. By an 'appropriate' point, we mean of course one which enables us to extend immediately our theory of integer displacements along the line to a theory of *rational* displacements involving the use of rational points and rational distances in the obvious way. Such a theory of displacements will then illustrate the additive structure of the rational numbers, just as our previous theory illustrated the structure of the integers.

To obtain the point $\frac{m}{n}$ we proceed as follows. Given the number line drawn in the plane, we use our ruler and take any other line through the point O (which represents the

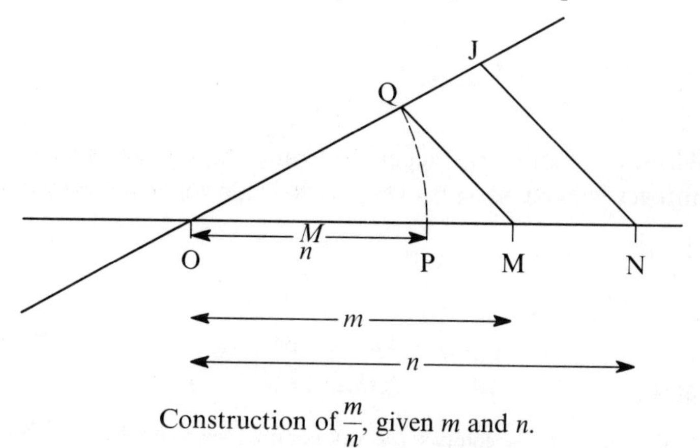

Construction of $\frac{m}{n}$, given m and n.

zero, 0). On this line we use our compasses and mark off the length 1. Suppose the point so marked is J, and the points on the number line corresponding to m, n are M, N respectively. Join NJ and construct† the line through M, parallel to NJ. Let this line cut OJ at Q. Mark off OP on the number line, so that OP = OQ. Since OQM and OJN are similar triangles, $OQ = \frac{OQ}{OJ} = \frac{OM}{ON} = \frac{m}{n}$, as required.

Returning to our algebraic theory, we must emphasize that despite its evident advantage over the integers, the rational number system is still neither large enough nor rich enough in structure for our algebraic purposes. If we stop at this point, restricting our work to the rational numbers, there will be basic algebraic processes which we shall be unable to perform. Thus, for example, although we can find rational solutions, $x = \pm 2$, and $x = \pm\frac{1}{4}$ respectively, for equations such as $x^2 = 4$ and $x^2 = \frac{1}{16}$, we cannot find rational solutions for equations such as $x^2 = 2$ or $3x^2 = 1$. Indeed, consider the first of these equations, namely $x^2 = 2$, and suppose we could find a rational solution representable in fractional form $x = \frac{m}{n}$. If we can find such a solution, we can assume that it is 'in its lowest terms', i.e. the integers m and n can be assumed to have no common factor. We then have

$$\left(\frac{m}{n}\right)^2 = 2$$

i.e.

$$m^2 = 2n^2.$$

Thus m^2 is an even integer; but only the square of an even integer is even, so m is even, i.e. $m = 2p$ for some integer p.

Thus

$$(2p)^2 = 2n^2,$$

so

$$4p^2 = 2n^2,$$

and

$$2p^2 = n^2.$$

† Do you know the compass and rule construction to do this?

We now use the same reasoning again, deducing from this equation that n must be even, i.e. $n = 2q$ for some integer q. But then m and n have a common factor 2, which contradicts our assumption that they had no common factor. But this was a valid assumption if we were correct in supposing in the first place that there could be a rational solution of the equation. This first supposition must therefore be false, since it has led to a contradiction; thus the equation $x^2 = 2$ cannot have a rational solution.

It is perhaps worth noting that this proof is entirely classical and was certainly known to the Greeks. In particular it employs a common form of logical argument, namely *proof by contradiction.* In such a proof we assume the result we wish to prove is false and deduce a contradiction; this is an extremely common logical technique and is one with which all would-be mathematicians should be conversant.

What we have really proved here is that we must extend our number systems even further if we wish to be able to find solutions of equations such as $x^2 = 2$. The appropriate extension is the *real number system.* Although we could once again proceed in a constructive fashion, observing how the real numbers can be built up from the rational numbers just as the rational numbers can be built up from the integers, we shall not do so. This is because the construction required is necessarily an infinite one in which we must use 'limiting' processes which take us beyond our finite algebraic processes. The reason for this becomes clearer if we recall how we approximate real numbers such as $\sqrt{2}$ by decimal expressions, correct to any number of 'places'. Thus, to one place of decimals, $\sqrt{2} \simeq 1{\cdot}4 = \frac{14}{10}$; to two places $\sqrt{2} \simeq 1{\cdot}41 = \frac{141}{100}$; to three places, $\sqrt{2} \simeq 1{\cdot}414 = \frac{1414}{1000}$; to four places $\sqrt{2} \simeq 1{\cdot}4139 = \frac{14\,139}{10\,000}$; etc. The real number $\sqrt{2}$ is the 'limit' of the *infinite* sequence of rational numbers $\frac{14}{10}, \frac{141}{100}, \frac{1414}{1000}, \ldots$. Thus whereas in constructing rational numbers we could proceed through a strictly finite algebraic process, using only pairs of integers to obtain fractional representations for rational numbers, here in the case of the real numbers we have no choice but to represent our real numbers, in some form or other, by infinite 'approximating sequences' of rational numbers.

We avoid this procedure, and the non-algebraic work involved, by the simple although possibly drastic expedient of once and for all stating a set of rules for the real numbers and asserting that these are the rules which completely determine the real number system. This we do in the next section. When we come to construct the complex numbers in the following chapter, it is this set of rules for the real numbers which forms the algebraic foundation for our work.

Worked examples 1.2

1 Deduce the statement 'given rational numbers a, b if $ab = 0$ then $a = 0$ or $b = 0$' from the following properties of rational numbers:

(i) multiplication of rational numbers is associative;
(ii) the rational number 1 acts as a multiplicative identity for the rationals;
(iii) any non-zero rational a has a multiplicative inverse a^{-1} such that $a^{-1}a = 1$;
(iv) $a0 = 0$ for any rational a.

Suppose a, b are any rationals such that $ab = 0$ and $a \neq 0$. Then by property (iii) a has a multiplicative inverse a^{-1}, thus $ab = 0$ implies $(a^{-1})(ab) = (a^{-1})0 = 0$ by property (iv). But then

$$
\begin{aligned}
0 &= (a^{-1})(ab) \\
&= (a^{-1}a)b && \text{(property (i))} \\
&= 1b && \text{(property (iii))} \\
&= b && \text{(property (ii))}.
\end{aligned}
$$

Thus, if $a \neq 0$, then $ab = 0$ implies $b = 0$, which proves the result.

2 Symbols 0, 1, 2, 3, 4 are multiplied by the rules embodied in the table

$\times$	0	1	2	3	4
0	0	0	0	0	0
1	0	1	2	3	4
2	0	2	4	1	3
3	0	3	1	4	2
4	0	4	3	2	1

so that, for example, $2 \times 3 = 3 \times 2 = 1$, and $3 \times 4 = 4 \times 3 = 2$, etc. Are there multiplicative inverses in this system?

We must first, if possible, find a multiplicative identity. Inspection of the table yields

$$\begin{aligned} 1 \times 0 &= 0 = 0 \times 1 \\ 1 \times 1 &= 1 = 1 \times 1 \\ 1 \times 2 &= 2 = 2 \times 1 \\ 1 \times 3 &= 3 = 3 \times 1 \\ 1 \times 4 &= 4 = 4 \times 1. \end{aligned}$$

Thus 1 is such an identity. By inspection of the table it is unique.

Inverses can now be read off, for

$$\begin{aligned} 1 \times 1 &= 1 = 1 \times 1 \\ 2 \times 3 &= 1 = 3 \times 2 \\ 4 \times 4 &= 1 = 4 \times 4 \end{aligned}$$

so that 1 is its own inverse, 3 is inverse to 2, 2 to 3 and 4 to itself. Again by inspection of the table, they are unique.

Exercises 1.2

1 Deduce from the properties of the integers given on page 24 that multiplication of rationals is (i) commutative, (ii) associative, (iii) distributive over addition of rationals.

2 Integers are combined by two rules, $\oplus$ and $\otimes$, given by

$$\begin{aligned} m \oplus n &= 0 \text{ if } m+n \text{ is even,} \\ &= 1 \text{ if } m+n \text{ is odd;} \\ m \otimes n &= 0 \text{ if } mn \text{ is even,} \\ &= 1 \text{ if } mn \text{ is odd.} \end{aligned}$$

What properties do the rules $\oplus$ and $\otimes$ have in common with ordinary addition and multiplication of integers? Is 'division' always possible?

3 If we replace 'integers' by 'rationals' in Worked examples 3 and 4 in section 1.1, will the proofs, and hence the results, remain valid?

1.3 Axioms for the real number field R

In our examination of the algebraic structure of the integers and the rationals, we have made no claim that the rules we obtained by extrapolation from our arithmetical experience were in any sense complete sets of rules for these systems of numbers. Our intention has been to introduce the reader to some elementary ideas about algebraic structure, and our treatment was not in any sense intended to be exhaustive. We were content to observe that our rules were reasonable in terms of our experience with numbers, and that many acceptable algebraic computations could be logically based upon them.

Now, in treating the real numbers, we do intend to give a complete set of rules. These will be stated not as observations from our experience (although obviously they must tally with our experience if they are to be acceptable), but as axioms, i.e. basic assumptions, on which the whole of the rest of our work can be logically based. The logically-minded reader will therefore probably wish to regard this as the true starting point of our work, and will be helped in taking this view by our statement of the fact (which we shall not prove!) that any kind of numbers satisfying our axioms will be algebraically indistinguishable from the real numbers.

First, then, we assert that the real numbers, which we denote by **R**, have the algebraic structure of a *field*, by which we mean there are two rules of combination for real numbers, addition and multiplication, and these two rules of combination satisfy the following *axioms for a field.*

I. *Additively*, we have

(i) for all a, b in **R**, $a+b$ is unique in **R**;
(ii) for all a, b in **R**, $a+b = b+a$;
(iii) for all a, b, c in **R**, $(a+b)+c = a+(b+c)$;
(iv) **R** contains a unique element 0 such that for all a in **R**,

$$a+0 = 0+a = a;$$

(v) for each a in **R**, there exists a unique element $(-a)$ in **R** such that

$$a+(-a) = 0 = (-a)+a.$$

II. *Multiplicatively* we have a system of very similar axioms:

(i) for all a, b in **R**, ab is unique in **R**;
(ii) for all a, b in **R**, $ab = ba$;
(iii) for all a, b, c in **R**, $a(bc) = (ab)c$;
(iv) **R** contains a unique element 1 $(\neq 0)$ such that for all a in **R**, $a1 = a = 1a$;
(v) for each $a \neq 0$ in **R** there exists a unique element a^{-1} in **R** such that

$$a(a^{-1}) = 1 = (a^{-1})a.$$

III. *Relating addition and multiplication* we have the axioms: for all a, b, c in **R**,

(i) $a(b+c) = ab+ac$;
(ii) $(a+b)c = ac+bc$.

Briefly, in I we observe that **R** is closed under addition and, also additively, is *commutative*, *associative*, has an *additive identity* 0, and has *additive inverses* $(-a)$ for all a in

R. In II we observe that **R** is closed under multiplication and, also multiplicatively, **R** is *commutative*, *associative*, has a *multiplicative identity* 1, and has *multiplicative inverses* a^{-1} for all non-zero elements a in **R**. Finally in III we link the two operations of addition and multiplication by two *distributive* rules. Once again we have restricted ourselves to the 'simplest cases' of associativity and commutativity. The reader is again referred to Appendix 1 for a justification of the assumption that these simplest cases imply the general cases.

The axioms we have given must be satisfied by any algebraic system if it is to be said to have the structure of a field. Our observations in section 1.1 imply that the integers do not constitute a field, since they do not satisfy the requirement of II(v) that all non-zero numbers in a field must have multiplicative inverses. So our axioms distinguish algebraically the real numbers from the integers. On the other hand, our observations about the rationals in section 1.2 imply that the rationals do satisfy the field axioms. Thus both the real and the rational numbers are examples of fields and in this sense as yet we have no algebraic rules enabling us to distinguish them from one another.

We next assert that the field **R** is an *ordered* field, i.e. we can speak of one real number being greater or less than another. We obtain this ordering by asserting that **R** contains a collection of numbers denoted by $\mathbf{R}^+$ and called the (strictly) *positive* real numbers, satisfying the following rules:

(i) for all a, b in $\mathbf{R}^+$, $a+b$ is in $\mathbf{R}^+$, and ab is in $\mathbf{R}^+$;

(ii) for any a in **R**, one and only one of the following relations holds: a is in $\mathbf{R}^+$, $a = 0$, or $(-a)$ is in $\mathbf{R}^+$.

Given the set of positive real numbers, $\mathbf{R}^+$, we define a real number a to be *less than* a real number b if $b-a$ $(= b+(-a))$ is positive, i.e. in $\mathbf{R}^+$. In this case we write $a < b$. Similarly, we define a to be *greater than* b, and write $a > b$, if $a-b$ is in $\mathbf{R}^+$. The notation $a \leqslant b$ is reserved for the case when $a < b$ or $a = b$, and similarly $a \geqslant b$ means that either $a > b$ or $a = b$. With these definitions, all the traditional theory of real inequalities becomes possible. In particular, for example, it follows that $x^2 > 0$ for any real number, and that $xy > 0$ if and only if either $x > 0$ and $y > 0$, or $x < 0$ and $y < 0$.

Although we have not stated these facts in sections 1.1 and 1.2, it is also true that the integers and the rational numbers can be ordered. Indeed we implicitly assume this in our pictorial illustration on the real number line when we mark off the positive integers and rationals to the right of the origin, and the greater of two numbers to the right of the lesser. Thus the real numbers are not to be distinguished from the rational numbers by reference to the ordering of the reals: both the rationals and real numbers are examples of ordered fields.

To distinguish the real numbers from the rational numbers in fact we require only one further axiom. To state this we need some terminology. First, then, given a non-empty set of numbers S in an ordered field, we say that a number b in the field is a *lower bound* of S, or that S is *bounded below by* b, if $b \leqslant x$ for all x in S. Given a set S which is bounded below, we say that a number in the field is a *greatest lower bound* of S if it is a lower bound for S and moreover greater than, or equal to, all other lower bounds. Finally we say that an ordered field is *complete* if, within it, every non-empty set of numbers which is bounded below has a greatest lower bound.

We can now state our last assumption about the field **R**.

Axiom. *The real numbers form a complete ordered field.*

On this all-embracing axiom the whole algebraic theory of real numbers can be made to rest. In particular, the real numbers as constructed from the rationals by means of the limiting processes mentioned above can be proved to satisfy this axiom. However, even more is true: any complete ordered field can be proved to be algebraically indistinguishable from the real numbers. In other words, the real numbers are algebraically 'characterized' by this axiom:

algebraically speaking, there is only one complete ordered field, namely the real numbers.

We shall not attempt to prove these facts. Neither shall we attempt, in what now follows, to give a complete logical justification for the various facts we observe which are consequences of our axiom. The proofs of these facts lie in the subject of *real analysis* with which we are not concerned here. Of course when these facts are used in our discussion of the complex numbers they can if necessary be regarded as further assumptions about the real numbers. Their logical deduction from our axiom is a natural part of any full treatment of the real field; here we shall move forward as quickly as possible to consider the field of complex numbers.

Let us first note then that we can regard the rationals and the integers as given to us, once we have defined our real numbers. This assertion depends in the first instance upon the observation that in the real field the numbers 0, 1, $1+1$, $1+1+1$, ..., and their additive inverses, are all distinct, and indistinguishable in their algebraic behaviour from the integers. We denote them, and their negatives, in the usual way by 0, ± 1, ± 2, ..., etc.

Given these 'integers' as part of the real field, we observe that all non-zero real numbers have multiplicative inverses and that in consequence within the real field we can derive from our integers the set of 'fractions' $\frac{1}{2} = (1)(2^{-1})$, $\frac{2}{3} = (2)(3^{-1})$, etc. Elementary consideration of how these fractions add, multiply, and are equal, then leads us to observe that they form an algebraic system indistinguishable from the rational numbers.

We next note without proof that the assumption of completeness of the real field finally distinguishes the real numbers from the rational numbers. In the rational field there are non-empty sets of numbers which are bounded below but for which there is no greatest (rational) lower bound. An example of such a set is that of all rational numbers x such that $x^2 > 2$; the point being that the number we might hope would be the greatest lower bound of this set is $\sqrt{2}$ which as we have seen is not a *rational* number, so is not available to us in the rational field. We may contrast this with the situation which occurs in the real field when we consider the set of all *real* numbers x such that $x^2 > 2$. Here we do have a greatest lower bound, namely the *real* number $\sqrt{2}$.

Again without proof we note that our use of *lower* bounds and *greatest* lower bounds should not be taken to imply that there is an unsymmetrical situation here. Sets of numbers can be bounded above, i.e. have *upper* bounds, and we can contemplate the existence of *least* upper bounds. In a complete field, defined in terms of lower bounds and greatest lower bounds, a non-empty set with an upper bound has in fact a least upper bound. Conversely we could define completeness in terms of upper bounds and least upper bounds. In such a case a non-empty set with a lower bound can be proved to have a greatest lower bound.

Finally we note for particular later use one further consequence of our completeness axiom, namely that if r is any real number, then there exists a unique positive real number, called the positive square root of r, denoted by $\sqrt{r}$ and such that $(\sqrt{r})^2 = r$. The proof of this fact, i.e. its strict logical deduction from our axioms for the real field, is given in Appendix 2. Here we shall observe that it can perhaps become more readily acceptable when seen in geometrical terms.

Once again therefore we turn to our number line. This line, which we have previously marked first with integers and then with rationals, is now to be regarded as completely marked off, in the sense that *on the (real) number line to every point there corresponds a real number and to every real number there corresponds a point.* This assertion is the geometrical analogue for our completeness axiom for the real field; what we are saying is that we must now assume that in our geometrical work we can mark off the points on the line using real numbers so that there are no 'gaps' on the line or in the real number field, i.e. no points are left unmarked and no numbers are excluded. Having made this assumption we immediately extend our theory of integral and rational displacements, so that movements and displacements on the number line can now be associated with real points and real distances, which need not be integral or rational. In all that follows we shall take this for granted. Here we are now concerned to see how this assumption guarantees that given a point on the line marked r, where $r > 0$, there is also a point marked '$\sqrt{r}$', which corresponds to the positive root of the equation $x^2 = r$. We give a geometrical construction to justify our belief that this is true.

Consider therefore the number line, lying as usual in the real plane. Denote the points representing the real numbers $-1, 0, 1$ and r by I′, O, I and R respectively. First construct† the circle with diameter I′R. Next erect† a perpendicular to the number line, through O. Let this perpendicular meet

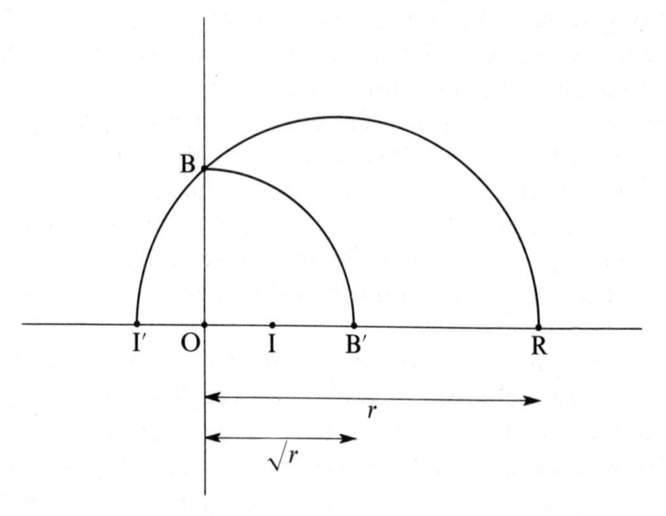

the circle at B. By standard Euclidean properties of the circle, $OB^2 = I'O.OR = OR = r$. Lay off the length OB in a positive direction along the number line, starting at O and finishing at B′; we then have the point B′ corresponding to '$\sqrt{r}$', as required.

†Do you know how to do this using only compass and ruler?

Of course we have to admit that such a 'construction' does not constitute a logical proof, and that algebraically much more would be needed to prove the existence of $\sqrt{r}$ as a consequence of our axioms for the real field. Nevertheless we may surely call on such a construction as support for our confidence in the truth of our assertion about the the existence of real square roots.

As another example of the way in which the geometrical theory of real displacements can and will help us in our algebraic work, we finish this chapter by considering the algebraic interpretation of the idea of the length of a displacement. 'Lengths' are, of course, non-negative numbers, and do not have a direction associated with them. In this, of course, they differ from displacements for when we speak of a displacement on the line we speak of its being of such and such a *length* and in such and such a *direction.* The geometrical association of a length with a displacement corresponds to associating algebraically with any real number x a non-negative real number, which must correspond to x in the same way that the length of a displacement corresponds to the displacement. In plain terms, we must 'drop the sign'. Let us therefore, given a real number x, define a non-negative real number which we denote by $|x|$, such that

$$\begin{aligned} |x| &= x \quad \text{if} \quad x \geqslant 0 \\ &= -x \quad \text{if} \quad x < 0. \end{aligned}$$

We call this number $|x|$ the *absolute value* of x. Geometrically interpreted $|x|$ is the length of the movement from the origin to the point representing the real number x; perhaps more usefully, $|x-a|$ can be interpreted as the distance from the point x on the real line to the point a.

In practical terms we now have available simple geometrical interpretations of inequalities such as $|x-1| < 2$, which can be illustrated by the set of points on the real line whose distance from the point 1 is less than 2. Algebraically of course this is the 'interval' $-1 < x < 3$. More generally the set of x such that $|x-a| < b$ is the set of x such that

$a-b < x < a+b$. Such a statement is surely self-evident geometrically when read in terms of lengths or distances. In algebraic terms it follows (logically, but perhaps less readily?) from our definition of the absolute value and our rules for ordering the real field, since if $|x-a| < b$, then we have both

$$x-a < b \quad \text{and} \quad -(x-a) < b,$$

in other words, both $b-(x-a)$ and $b+(x-a)$ are positive. But $b-(x-a) = (a+b)-x$, thus $a+b > x$; also $b+(x-a) = x-(a-b)$, thus $x > a-b$. All in all, we write

$$a-b < x < a+b.$$

In particular if $a = 0$, $|x| < b$ implies $-b < x < b$.

There are certain important inequalities concerning absolute values in relation to the operations of addition and subtraction in the field **R**. We close by giving these.

Theorem. *For any real numbers a, b,*

(i) $|a+b| \leqslant |a|+|b|$;

(ii) $|a-b| \geqslant \big||a|-|b|\big|$.

Proof.

(i) By definition of absolute value $|a|-a$ is positive or zero according as $a < 0$ or $a \geqslant 0$, and $a+|a|$ is similarly non-negative. Thus $|a| \geqslant a$ and $a \geqslant -|a|$. Similarly $|b| \geqslant b$ and $b \geqslant -|b|$.

Now inequalities can be added (for if $u > v$ and $s > t$ we have $u-v$ and $s-t$ positive so $(u-v)+(s-t) = (u+s)-(v+t)$ is positive, i.e. $u+s > v+t$). Thus adding the above inequalities we have $|a|+|b| \geqslant a+b$ and $a+b \geqslant -(|a|+|b|)$. But these two inequalities together yield

$$-(|a|+|b|) \leqslant a+b \leqslant |a|+|b|,$$

i.e.

$$|a+b| \leqslant |a|+|b|.$$

(ii) We have

$$\begin{aligned} |a| &= |a-b+b| \\ &\leqslant |a-b|+|b| \qquad \text{(by (i))}. \end{aligned}$$

Thus $|a|-(|b|+|a-b|) = (|a|-|b|)-|a-b|$ is not positive, i.e.

$$|a|-|b| \leqslant |a-b|.$$

Similarly
$$|b| = |b-a+a|$$
$$\leqslant |b-a|+|a| \qquad \text{(by (i))}.$$

Thus
$$|b|-|a| \leqslant |b-a|$$
$$= |a-b|,$$

or
$$|a|-|b| \geqslant -|a-b|,$$

for if $x \leqslant y$ then $-y \leqslant -x$.

Taking both inequalities together, we have

$$-|a-b| \leqslant |a|-|b| \leqslant |a-b|,$$

whence
$$||a|-|b|| \leqslant |a-b|,$$

as required.

Worked examples 1.3

1 Deduce from the axioms I, II and III for a field, on page 45,
(i) that $0a = 0$ for all a in a field,
(ii) that $(-x)(-y) = xy$ for all x, y in a field.

(i) $(0+a)a = aa$ (since $0+a = a$)

so $0a+aa = aa$ (by distributivity)

Adding $-aa$ to both sides of this equation, and using associativity of addition, we find immediately that $0a = 0$ as required. (Cf Worked examples 1.1, question 3, p. 26.)
(ii) Again in Worked examples 1.1, question 4, we used rules for integers to prove

$$(m)(-n) = (-m)(n) = -(mn)$$

for all integers m, n. The rules we used hold in a field. The same argument therefore proves

$$(a)(-b) = (-a)(b) = -(ab)$$

for all a, b in a field.

Thus

$$(-x)(-y) = -(x(-y))$$

$$= -(-(xy))$$

But $$-(xy)+xy = 0,$$

so $$-(-(xy)) = xy.$$

2 Prove that the set of real numbers of the form $a+b\sqrt{2}$, where a and b are any rational numbers, combine under real addition and multiplication to form a field, i.e. they have properties I(i)–(v), II(i)–(v), and III(i) and (ii), on page 45.

We must first prove that numbers of the given kind combine under $+$ and $\times$ to give numbers of the same kind, i.e. that we have closure of the given set of numbers under $+$ and $\times$. We have

$$(a+b\sqrt{2})+(a'+b'\sqrt{2}) = (a+a')+(b+b')\sqrt{2}.$$

But if a, a', b and b' are rational, so also are $a+a'$ and $b+b'$; hence the set of numbers is closed under $+$. We also have

$$(a+b\sqrt{2})(a'+b'\sqrt{2}) = (aa'+2bb')+(ba'+ab')\sqrt{2},$$

so that closure under $\times$ follows, since $aa'+2bb'$ and $ba'+ab'$ are rational if a, a', b and b' are rational.

The real operations $+$ and $\times$ are both commutative, associative, and distributive in the real field. They must therefore remain so when restricted to the set of real numbers of the form $a+b\sqrt{2}$. To complete our proof we therefore only need to find additive and multiplicative identities and inverses in our set.

The additive identity is of course

$$0+0\sqrt{2} = 0,$$

and the multiplicative identity is

$$1+0\sqrt{2} = 1.$$

The appropriate properties of these identities follow immediately from the properties of 0 and 1.

Inverses are readily computed. Thus

$$-(a+b\sqrt{2}) = (-a)+(-b)\sqrt{2},$$

and $$(a+b\sqrt{2})^{-1} = \frac{a}{a^2-2b^2} - \frac{b}{a^2-2b^2}\sqrt{2}.$$

The second of these is perhaps less obvious than the first, but follows from the fact that in the field of real numbers,

$$\frac{1}{a+b\sqrt{2}} = \frac{a-b\sqrt{2}}{(a+b\sqrt{2})(a-b\sqrt{2})}$$

$$= \frac{a-b\sqrt{2}}{a^2-2b^2}$$

Notice that, as required for these computations, if a and b are rational, $a^2-2b^2 \neq 0$.

3 Deduce the following rules for inequalities from the axioms for an ordered field.

(i) For all $x \neq 0$, $x^2 > 0$,
(ii) if $a > b$ and $c > 0$, then $ac > bc$,
(iii) if $a > b$ and $c < 0$, then $ac < bc$,
(iv) if $x > 0$, then $x^{-1} > 0$,
(v) $xy > 0$ implies either $x > 0$ and $y > 0$ or $x < 0$ and $y < 0$,
(vi) $x^2 > y^2$ if and only if $|x| > |y|$.

(i) If $x \neq 0$, then $x > 0$ or $x < 0$. If $x > 0$, then it follows from the axioms for an ordered field that $x^2 = xx > 0$. If $x < 0$, then $-x > 0$ and $(-x)^2 > 0$. But it follows from Worked example 1(ii) above, that $x^2 = xx = (-x)(-x) = (-x)^2$; so again $x^2 > 0$.

(ii) It follows from our rules for ordering that $a > b$ implies $a-b > 0$. If $c > 0$ also, it follows again from our rules for an ordered field, that $(a-b)c > 0$. But then $ac-bc > 0$ by distributivity, i.e. $ac > bc$.

(iii) Again if $a > b$, then $a-b > 0$. Also if $c < 0$, then $-c > 0$, so that $a > b$, $c < 0$ imply $(a-b)(-c) > 0$, i.e. by Worked example 1(ii) above, and by distributivity, $-ac+bc > 0$, whence $ac < bc$.

(iv) Whatever $x \neq 0$ may be, either $x^{-1} > 0$ or $x^{-1} \leqslant 0$, but not both. If $x^{-1} \leqslant 0$ and $x > 0$, then $-x^{-1} \geqslant 0$ and $x > 0$ so that $(-x^{-1})(x) = -1 \geqslant 0$. But $1 = 1^2$ which is positive by (i) above. Thus $1 > 0$ and $-1 < 0$. We therefore have a contradiction and $x^{-1} > 0$ is implied by $x > 0$.

(v) If $xy > 0$, then $xy \neq 0$ and $x \neq 0$. Thus $x > 0$ or $x < 0$. If $x > 0$, then by (iv) above, $x^{-1} > 0$, which together with $xy > 0$ implies $(x^{-1})(xy) = (x^{-1}x)y = 1y = y > 0$. On the other hand, if $x < 0$, then $-x > 0$. But $xy = (-x)(-y) > 0$ so by the first part of this proof it follows that $-y > 0$, i.e. $y < 0$ as required.

(vi) Let $|x| > |y|$. Then by (ii), since $|y| > 0$, we have $|x|\,|y| > |y|^2$. But again by (ii), since $|x| > 0$, we then have $|y|^2 < |x|\,|y| < |x|^2$. Since $y^2 = |y|^2$ and $x^2 = |x|^2$, it follows that $y^2 < x^2$, as required.

Conversely, suppose $x^2 > y^2$. Then we cannot have $|y| \geqslant |x|$, for this, by the above argument with $\geqslant$ replacing $>$, would imply $y^2 \geqslant x^2$. So we must have $|x| > |y|$.

4 Find all real values of x such that
(i) $x^2-3x-4>0$,
(ii) $|x-3| \geqslant |2x-1|$
Illustrate your answers.

(i)

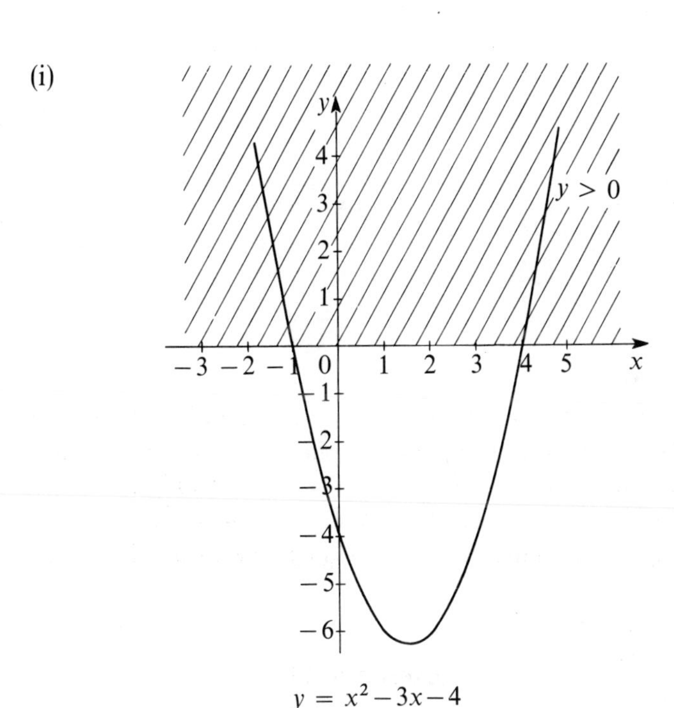

$y = x^2-3x-4$

Graphically we see that the inequality will hold for all $x>4$ and all $x<-1$. Algebraically we note that $(x-4)(x+1)>0$ if and only if the two factors are either both positive or both negative. We therefore require either $x>4$ and $x>-1$ or $x<4$ and $x<-1$. These two conditions are equivalent to $x>4$ and $x<-1$, since if $x>4$, certainly $x>-1$, and if $x<-1$, certainly $x<4$.

(ii)

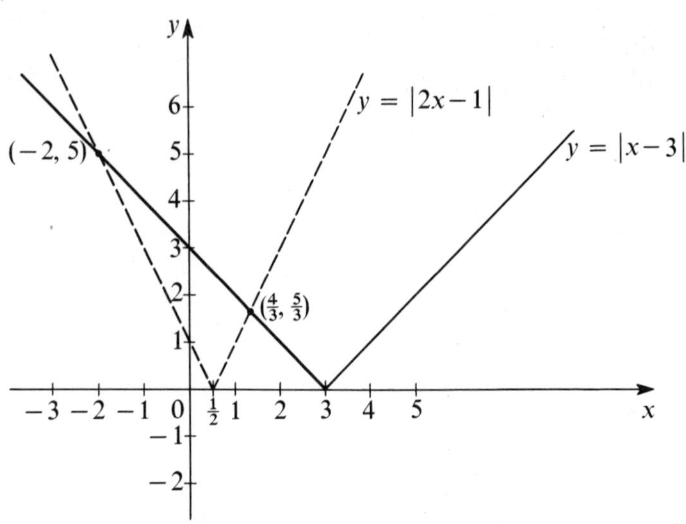

We have $|x-3| \geqslant |2x-1|$ if and only if $(x-3)^2 \geqslant (2x-1)^2$, in consequence of part (vi) of the preceding example. But $(x-3)^2 \geqslant (2x-1)^2$ if and only if $3x^2+2x-8 \leqslant 0$, i.e. if and only if $(3x-4)(x+2) \leqslant 0$. The inequality is therefore satisfied if and only if $-2 \leqslant x \leqslant \frac{4}{3}$.

Exercises 1.3

1 Assuming $\sqrt{3}$ is not rational, verify that the set of all real numbers of the form $a+b\sqrt{3}$, where a and b are any rational numbers, combine under ordinary real addition and multiplication to form a field.

2 Prove that a rational number $\dfrac{m}{n}$ is positive if and only if mn is positive.

3 Prove that in the real field $a^2-b^2 = (a-b)(a+b)$, and hence deduce that if $a^2 > b^2$ and $a, b > 0$, then $a > b$.

4 Find all real values of x such that

(i) $x > -1$ and $0 < \dfrac{\quad}{1+x^2} < 1$,

(ii) $\dfrac{1}{x-3} > 0$,

(iii) $x(1+x) < 1$,

(iv) $|x-2| > 3$,

(v) $|x+1| < |x|$.

Illustrate your answers.

Miscellaneous exercises 1

1 If a and b are any *integers*, does the set of real numbers of the form $a+b\sqrt{2}$ form a field?

2 Deduce from the rules of ordering that there is no real number a such that $a^2+1 = 0$.

3 Prove that there is no rational number $\dfrac{m}{n}$ such that $\dfrac{m^2}{n^2} = 3$.

(Hint: an integer is either of the form $3r$, $3r+1$, or $3r+2$; consider the divisibility by 3 of the squares of $3r+1$ and $3r+2$).

4 Rational numbers are hopefully combined in fractional form by setting $\dfrac{m}{n} * \dfrac{p}{q} = \dfrac{m+p}{n+q}$. Does such a rule define a satisfactory method of combining rational numbers?

5 Symbols 0, 1 are combined by the following rules for addition and multiplication:

$+$	0	1
0	0	1
1	1	0

$\times$	0	1
0	0	0
1	0	1

Prove that with these rules, the symbols 0 and 1 satisfy the axioms for a field. Can this field be ordered?

6 Deduce from the axioms for an ordered field that if a is a non-zero real number then

$$a+a+\ldots+a \neq 0,$$

whatever number of summands is taken.

7 Solve the inequalities

(i) $x(x+1)(x^2-9) < 0$,

(ii) $|x^3| > |8x|$,

(iii) $|2x-1|+|2x+1| > 8$,

(iv) $|x^2-1| < |2x|$.

8 Let X be a subset of the real numbers which is bounded above. Prove that a least upper bound for X is $-l$ where l is the greatest lower bound for the set $-X$ of elements $-x$ for x in X.

CHAPTER TWO

The two-dimensional real vector space $\mathbf{R}^2$

2.1 Construction of $\mathbf{R}^2$

We have already observed in the Introduction that complex numbers are formed from pairs of real numbers. Thus, having given the complete set of algebraic rules for the real number field $\mathbf{R}$, we naturally now turn to consider the rules governing the algebraic behaviour of pairs of such numbers.

Once again we shall start with an informal discussion, before giving formal definitions and rules. The corresponding discussion in Chapter One was algebraic in character and used our arithmetic understanding of integers and rational numbers as a guide to appropriate algebraic rules for real numbers. Here, on the other hand, we start with a geometric discussion and use our knowledge of two-dimensional co-ordinate geometry to develop a geometric theory of displacements in the plane as a guide to the appropriate algebraic rules for pairs of real numbers.

The theory of displacements in a plane which we seek is a two-dimensional generalization of the one-dimensional theory which we used to illustrate the additive structure of the integers, the rationals and the real number field $\mathbf{R}$ in the last chapter. We may recall that in that chapter we marked points or 'places' on the real line by means of real 'co-ordinates' and then interpreted our numbers pictorially by means of movements on the line from one 'place' to another: movements represented displacements, the former being tied to particular starting points, the latter not. Taking the real plane in place of the real line, we now mark points in it in the usual plane co-ordinate geometric way in relation to an origin O and perpendicular axes $0x$, $0y$, so that each 'place' in the plane is marked with two real co-ordinates. Now, given a starting place in the plane, a movement from it is determined by a pair of real numbers, the first fixing the amount of movement 'horizontally', the second the amount to be moved 'vertically'. Once again, movements represent displacements. For the latter we require only a 'horizontal component' and a 'vertical component' and no starting point. Both movements and displacements require a pair of real numbers to determine them, but a movement is tied to a particular starting point, a displacement is not.

A pair of real numbers can thus be given two geometric interpretations; one as the co-ordinates of a point in the real

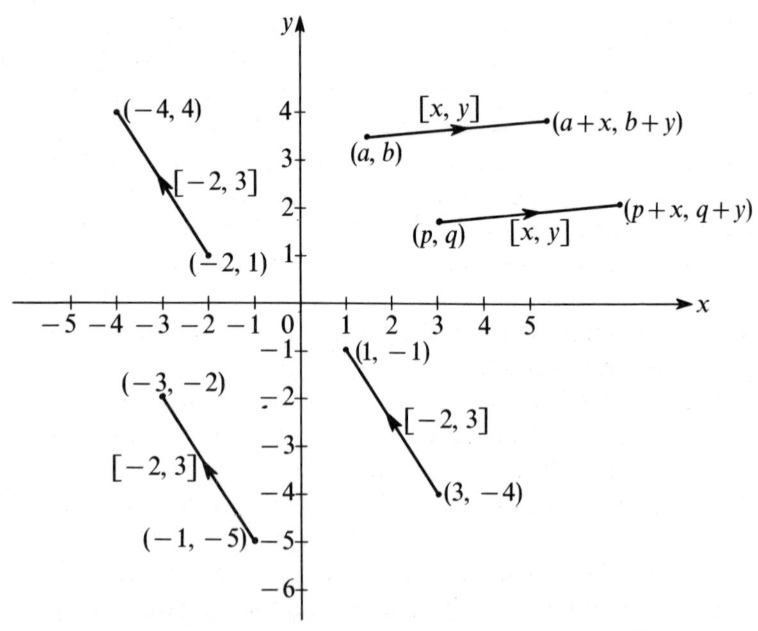

Examples of displacements and movements

plane, the other as a displacement and hence as a movement in the plane. To begin with at least, we shall find it convenient to distinguish notationally between pairs of real numbers in relation to these two interpretations, and in consequence we use round brackets (x, y) to denote the point in the plane with co-ordinates x and y, and square brackets $[x, y]$ to denote the displacement determined by moving a distance $|x|$ horizontally, to the right or left according as x is positive or negative, and $|y|$ vertically, up or down according as y is positive or negative. Given a starting point with co-ordinates (a, b), a displacement $[x, y]$ can thus be represented by a movement from (a, b) to $(a+x, b+y)$. Conversely, given points $P(x_P, y_P)$ and $Q(x_Q, y_Q)$, a move from P to Q represents a displacement $[x_Q - x_P, y_Q - y_P]$.

Just as in the case of displacements on the real line, in dealing geometrically with displacements or movements in the plane, it is natural to consider the effect of following one displacement, or movement, by another. Thus for example we may move from a point A in the plane to a point B, and then from the point B to a point C, in which case, putting

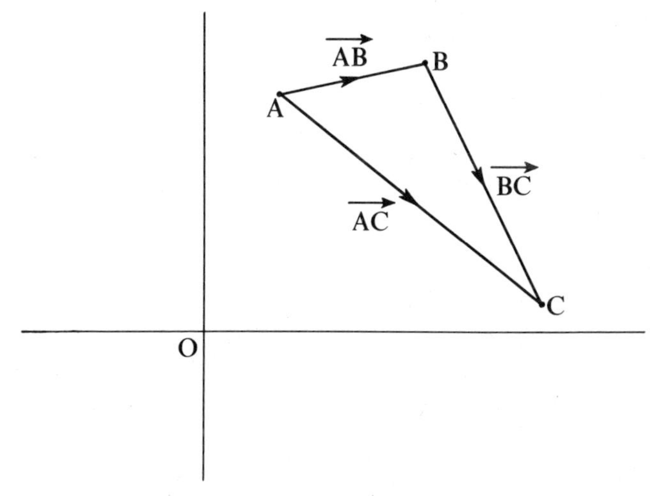

Composition of displacements illustrated by composition of movements

the two movements together, or 'composing' them as we again say, we have *in toto* a move from A to C. If we use the notation $\overrightarrow{AB}$ to denote the displacement represented by the move from A to B, and $\overrightarrow{BC}$, $\overrightarrow{AC}$ in a similar fashion in relation to the moves from B to C and A to C respectively, we can write the relationship between $\overrightarrow{AB}$, $\overrightarrow{BC}$ and $\overrightarrow{AC}$ in a pleasing algebraic additive form

$$\overrightarrow{AB}+\overrightarrow{BC} = \overrightarrow{AC}.$$

That we find it appropriate to speak of 'adding' displacements is even more reasonable when we consider the addition of two displacements given in the form $[x_1, y_1]$ and $[x_2, y_2]$. If we compose these displacements, representing them respectively by movements from (a, b) to $(a+x_1, b+y_1)$ and then from $(a+x_1, b+y_1)$ to $(a+x_1+x_2, b+y_1+y_2)$, we move in effect from (a, b) to $(a+x_1+x_2, b+y_1+y_2)$. In other words, displacements $[x_1, y_1]$ and $[x_2, y_2]$ are added by the formula

$$[x_1, y_1]+[x_2, y_2] = [x_1+x_2, y_1+y_2],$$

i.e. we add the displacements by adding each 'component' of the displacements separately, using the usual rules for adding real numbers.

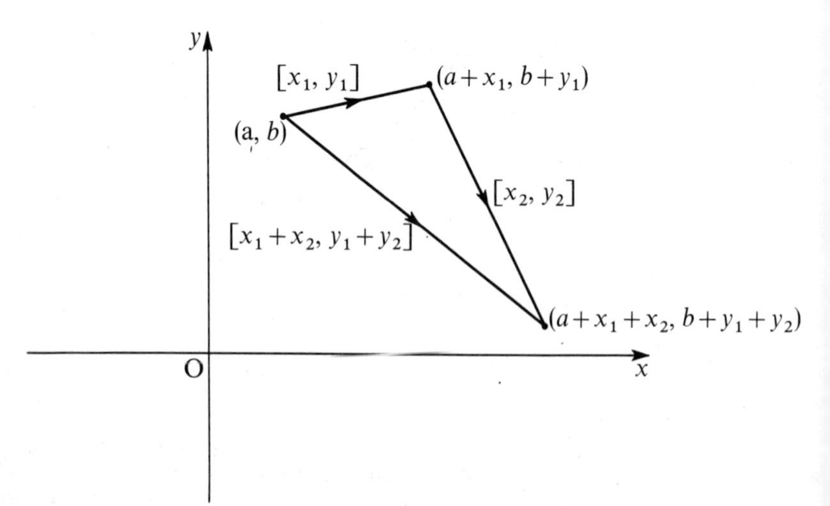

The reader who is familiar with the elementary theory of forces in applied mathematics will recognize that our rule for adding displacements is similar to the 'triangle of forces' rule which is used to obtain the resultant of two forces. In fact both displacements and forces are examples of mathematical objects with which are associated both a 'magnitude', or 'size', and a 'direction of action'. Any such objects are traditionally called *vectors*. Their algebraic manipulation, called *vector algebra*, is the same, whether the vectors concerned are force vectors, displacement vectors, or any other kind of vectors.

In elementary force theory we also have a 'parallelogram of forces' rule which is used to obtain the resultant of two forces. There is a similar rule in displacement theory which is worth considering in relation to the rules of algebra we expect our vectors to satisfy. Thus if $[x_1, y_1]$ and $[x_2, y_2]$ are illustrated by movements from P(a, b) to Q$(a+x_1, b+y_1)$ and S$(a+x_2, b+y_2)$ respectively, then their sum $[x_1+x_2, y_1+y_2]$ can be represented by a move from P(a, b) to the point R, with co-ordinates $(a+x_1+x_2, b+y_1+y_2)$ which completes the parallelogram PQRS. In case P is the origin O, when $(a, b) = (0, 0)$, we have the following, perhaps most convenient picture.

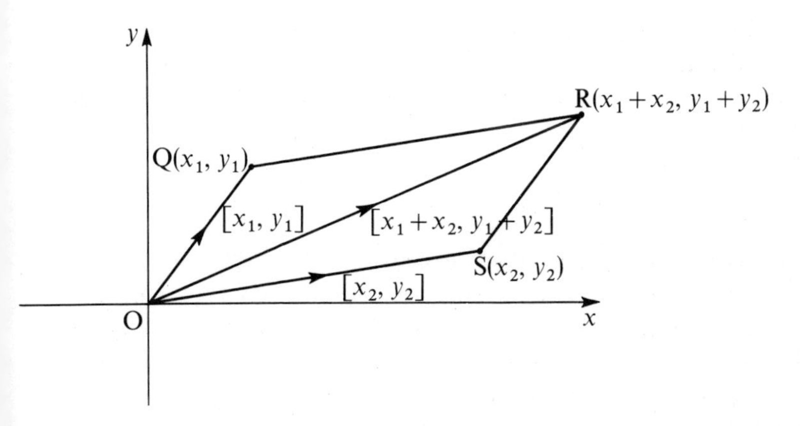

Thus, just as a parallelogram of forces consists of two triangles of forces, so also in this picture we have two triangles illustrating the composition of $[x_1, y_1]$ and $[x_2, y_2]$. One, the triangle OQR, is associated with $\overrightarrow{OQ}+\overrightarrow{QR}$; the other, triangle OSR, is associated with the composition $\overrightarrow{OS}+\overrightarrow{SR}$. But $\overrightarrow{OQ} = \overrightarrow{SR} = [x_1, y_1]$, $\overrightarrow{QR} = \overrightarrow{OS} = [x_2, y_2]$, and $\overrightarrow{OQ}+\overrightarrow{QR} = \overrightarrow{OR} = \overrightarrow{OS}+\overrightarrow{SR}$. We have thus built into our parallelogram rule an algebraic rule for commutativity of composition of displacement vectors:

$$\overrightarrow{SR}+\overrightarrow{OS} = \overrightarrow{OS}+\overrightarrow{SR},$$

since both compositions yield the vector $\overrightarrow{OR}$.

Taking a geometric point of view one might therefore assert that the basic geometric property of the parallelogram, i.e. the fact that opposite sides of a parallelogram are equal in length and parallel (which ensures that $\overrightarrow{OQ} = \overrightarrow{SR}$ and $\overrightarrow{OS} = \overrightarrow{QR}$) implies that composition of displacement vectors is commutative. On the other hand, from an algebraic point of view, writing our displacements in the form $[x_1, y_1]$ and $[x_2, y_2]$, we have

$$[x_1, y_1]+[x_2, y_2] = [x_1+x_2, y_1+y_2]$$

and $$[x_2, y_2]+[x_1, y_1] = [x_2+x_1, y_2+y_1].$$

Thus, algebraically, it appears that it is commutativity of real addition (namely the fact that $x_1+x_2 = x_2+x_1$ and $y_1+y_2 = y_2+y_1$), which implies that we have the commutative rule:

$$[x_1, y_1]+[x_2, y_2] = [x_2, y_2]+[x_1, y_1].$$

If we assert for either geometrical or algebraic reasons that commutativity holds for composition of displacements, what about associativity? Or for that matter what about any of the other rules of algebra which we have observed govern the addition of real numbers?

Consider associativity first. If three displacements are represented by movements from A to B, B to C and C to D respectively, we have

$$\overrightarrow{AB}+(\overrightarrow{BC}+\overrightarrow{CD}) = \overrightarrow{AB}+\overrightarrow{BD}$$
$$= \overrightarrow{AD},$$

and
$$(\overrightarrow{AB}+\overrightarrow{BC})+\overrightarrow{CD} = \overrightarrow{AC}+\overrightarrow{CD}$$
$$= \overrightarrow{AD}.$$

So taking a geometric view, we have an associative rule for composition of displacements. Algebraically, this follows from the associative property of real addition. Thus

$$\begin{aligned}[x_1, y_1]+([x_2, y_2]+[x_3, y_3]) \\ &= [x_1, y_1]+[x_2+x_3, y_2+y_3] \\ &= [x_1+(x_2+x_3), y_1+(y_2+y_3)] \\ &= [(x_1+x_2)+x_3, (y_1+y_2)+y_3] \\ &= [x_1+x_2, y_1+y_2]+[x_3, y_3] \\ &= ([x_1, y_1]+[x_2, y_2])+[x_3, y_3].\end{aligned}$$

Next, there is an obvious geometric (and algebraic) candidate for the 'zero' displacement, namely the displacement through no distance; algebraically, the displacement vector $[0, 0]$.

Finally, additive inverses exist, since associated with any displacement is the displacement of equal length in the opposite direction, which when composed with the given displacement leaves you where you were to start with, unmoved. In algebraic terms,

$$-[x, y] = [-x, -y],$$

since $[x, y]+[-x, -y] = [x+(-x), y+(-y)] = [0, 0]$.

It appears therefore, if our geometrical theory of addition of displacement vectors is to be regarded as an accurate guide, that addition of vectors in the abstract satisfies much the same rules as ordinary addition of numbers. The algebraic work we have carried out with the pairs $[x, y]$ would seem to confirm this and also imply that a proof of this fact could be made to rest entirely on our rules for addition of real numbers. To check this, we must drop our geometric displacement theory and proceed to consider pairs of real numbers in the abstract, by which we mean pairs of numbers just as pairs, without any interpretation placed upon them. To emphasise that we are no longer concerned with displacement vectors, we shall now write our pairs in the form (x, y). There will be no danger of confusing these pairs with the co-ordinates of points in the real plane.

Thus, let $\mathbf{R}^2$ denote the collection of all *ordered pairs* of real numbers (x, y). By ordered pairs we mean pairs of numbers (x, y) in which we distinguish between the *first* number, x, and the *second*, y. We shall refer to these pairs as *two-dimensional real vectors*. When we use a single letter to denote such a pair we shall use bold type, e.g. $\mathbf{u} = (x, y)$.

We first define equality of these vectors.

Definition. $(x, y) = (x', y')$ if and only if $x = x'$ and $y = y'$.

Next we define vector addition.

Definition. The sum of the vectors (x_1, y_1) and (x_2, y_2) is defined to be (x_1+x_2, y_1+y_2). We write

$$(x_1, y_1)+(x_2, y_2) = (x_1+x_2, y_1+y_2).$$

We now prove that vector addition is commutative, associative, has a 'zero', and has 'additive inverses', just like integer addition, rational addition or real addition.

Theorem. *If* $\mathbf{u}$, $\mathbf{v}$, $\mathbf{w}$ *are vectors in* $\mathbf{R}^2$, *with addition defined as above, then*

(i) $\mathbf{u}+\mathbf{v} = \mathbf{v}+\mathbf{u}$;

(ii) $(\mathbf{u}+\mathbf{v})+\mathbf{w} = \mathbf{u}+(\mathbf{v}+\mathbf{w})$;

(iii) *there is a unique 'zero' vector* $(0, 0)$ *in* $\mathbf{R}^2$ *denoted by* $\mathbf{0}$, *such that* $\mathbf{u}+\mathbf{0} = \mathbf{u} = \mathbf{0}+\mathbf{u}$;

(iv) *for each vector* $\mathbf{u} = (x, y)$ *in* $\mathbf{R}^2$, *there is a unique 'additive inverse' vector, written* $-\mathbf{u}$ *and given by* $-\mathbf{u} = (-x, -y)$, *such that*

$$\mathbf{u}+(-\mathbf{u}) = \mathbf{0} = (-\mathbf{u})+\mathbf{u}.$$

Proof. These properties we now deduce from our rules for the real field $\mathbf{R}$. Thus, if $\mathbf{u} = (x, y)$, and $\mathbf{v} = (x', y')$, where x, y, x', y' are any real numbers, we have

$$\begin{aligned}
\text{(i) } \mathbf{u}+\mathbf{v} &= (x, y)+(x', y') \\
&= (x+x', y+y') \\
&= (x'+x, y'+y) \quad \text{(by commutativity in } \mathbf{R}) \\
&= (x', y')+(x, y) \\
&= \mathbf{v}+\mathbf{u}.
\end{aligned}$$

If $w = (x'', y'')$, we have

$$\begin{aligned}
\text{(ii) } (\mathbf{u}+\mathbf{v})+\mathbf{w} &= ((x, y)+(x', y'))+(x'', y'') \\
&= (x+x', y+y')+(x'', y'') \\
&= ((x+x')+x'', (y+y')+y'') \\
&= (x+(x'+x''), y+(y'+y'')) \\
&\qquad\qquad \text{(by associativity in } \mathbf{R}) \\
&= (x\ y)+(x'+x'', y'+y'') \\
&= (x, y)+((x', y')+(x'', y'')) \\
&= \mathbf{u}+(\mathbf{v}+\mathbf{w}).
\end{aligned}$$

Again, we have

$$\begin{aligned}
\text{(iii) } \mathbf{u}+\mathbf{0} &= (x, y)+(0, 0) \\
&= (x+0, y+0) \\
&= (x, y) \qquad \text{(by definition of 0 in } \mathbf{R}) \\
&= \mathbf{u},
\end{aligned}$$

and similarly

$$\mathbf{0}+\mathbf{u} = \mathbf{u}.$$

That $\mathbf{0} = (0, 0)$ is the *unique* vector such that

$$\mathbf{u}+\mathbf{0} = \mathbf{u} = \mathbf{0}+\mathbf{u},$$

is an immediate consequence of the properties of the real number 0, for if a and b are such that

$$(x, y)+(a, b) = (x, y)$$

for all x, y, then

$$x+a = x$$

and

$$y+b = y$$

for all x, y. But our rules for the real field $\mathbf{R}$ then imply that $a = b = 0$, as required.

Finally, we have

(iv)

$$\begin{aligned}\mathbf{u}+(-\mathbf{u}) &= (x, y)+(-x, -y)\\ &= (x+(-x), y+(-y))\\ &= (0, 0)\\ &\quad \text{(by definition of } (-x) \text{ and } (-y) \text{ in } \mathbf{R})\\ &= \mathbf{0},\end{aligned}$$

and similarly

$$(-\mathbf{u})+(\mathbf{u}) = \mathbf{0}.$$

Again, $-\mathbf{u} = (-x, -y)$ is the *unique* vector with the property

$$\mathbf{u}+(-\mathbf{u}) = \mathbf{0},$$

for if

$$(x, y)+(a, b) = (0, 0)$$

then

$$x+a = 0$$

and

$$y+b = 0.$$

But these real equations imply $a = -x$ and $b = -y$, as required.

Thus we obtain logical confirmation that additive vector algebra is, in the abstract as well as in the particular examples of displacement vectors or force vectors, governed by rules which are identical with those governing addition of integers, rationals or real numbers in general. Additively, $\mathbf{R}$ and $\mathbf{R}^2$ have this algebraic structure in common. We can therefore deduce, for example, in the usual way†, that having defined the sum of any finite number of real two-dimensional vectors, such a sum is independent of the order in which the vectors are given, and of the order in which we perform the individual additions in the summation.

Of course, we do not discard our illustration of vectors by displacements in the plane. It still enables us to give helpful pictures of our algebraic processes. Thus, for example, if as before we define 'subtraction' in terms of 'addition of the inverse', so that $\mathbf{u}-\mathbf{v} = \mathbf{u}+(-\mathbf{v})$, we can illustrate the process involved by taking $\mathbf{u} = (x_1, y_1)$, $\mathbf{v} = (x_2, y_2)$ and noting that in displacement terms

$$\begin{aligned}[x_1, y_1]-[x_2, y_2] &= [x_1, y_1]+[-x_2, -y_2] \\ &= [x_1-x_2, y_1-y_2].\end{aligned}$$

We have the picture overleaf.

Given that we are still concerned to use our geometry of displacements for illustrative purposes in connection with addition of vectors, we may also ask if we can exploit this geometrical theory any further to our algebraic advantage. In one way we certainly can, for since a *length* is associated with any displacement, we can make a displacement longer or shorter (or indeed reverse its direction) by multiplying the length of the displacement by a real number. In terms of a particular movement say from O(0, 0) to P(x, y), representing a given displacement $\overrightarrow{OP}$, this would correspond to

† See Appendix 1. The proof there refers to numbers, but is obviously also applicable to vectors.

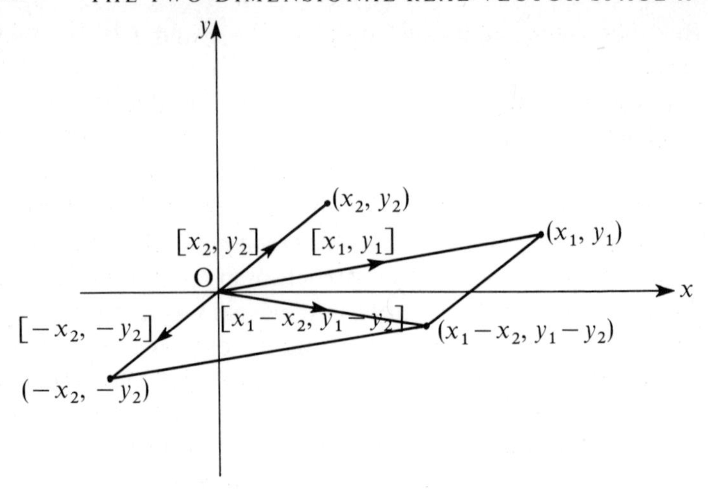

replacing $P(x, y)$ by $P'(\alpha x, \alpha y)$ where α is any real number. The diagrams opposite illustrate the various possibilities in case P is in the first quadrant.

The following definition and theorem make algebraic sense out of this geometrical process.

Definition. Given a real number α and an ordered pair (x, y), we define their product $\alpha(x, y)$ to be the pair $(\alpha x, \alpha y)$.

Theorem. *If* $\mathbf{u}$, $\mathbf{v}$ *are any ordered pairs of real numbers and* α, β *any real numbers, then*

$$\text{(v)} \quad \alpha(\mathbf{u}+\mathbf{v}) = \alpha\mathbf{u}+\alpha\mathbf{v},$$

$$(\alpha+\beta)\mathbf{u} = \alpha\mathbf{u}+\beta\mathbf{u};$$

$$\text{(vi)} \quad (\alpha\beta)\mathbf{u} = \alpha(\beta\mathbf{u});$$

$$\text{(vii)} \quad 1\mathbf{u} = \mathbf{u}.$$

Proof. The proofs of these results are again immediate consequences of the rules for $\mathbf{R}$. Thus if $\mathbf{u} = (x, y)$ and $\mathbf{v} = (x', y')$, then

$$\alpha(\mathbf{u}+\mathbf{v}) = \alpha((x, y)+(x', y'))$$

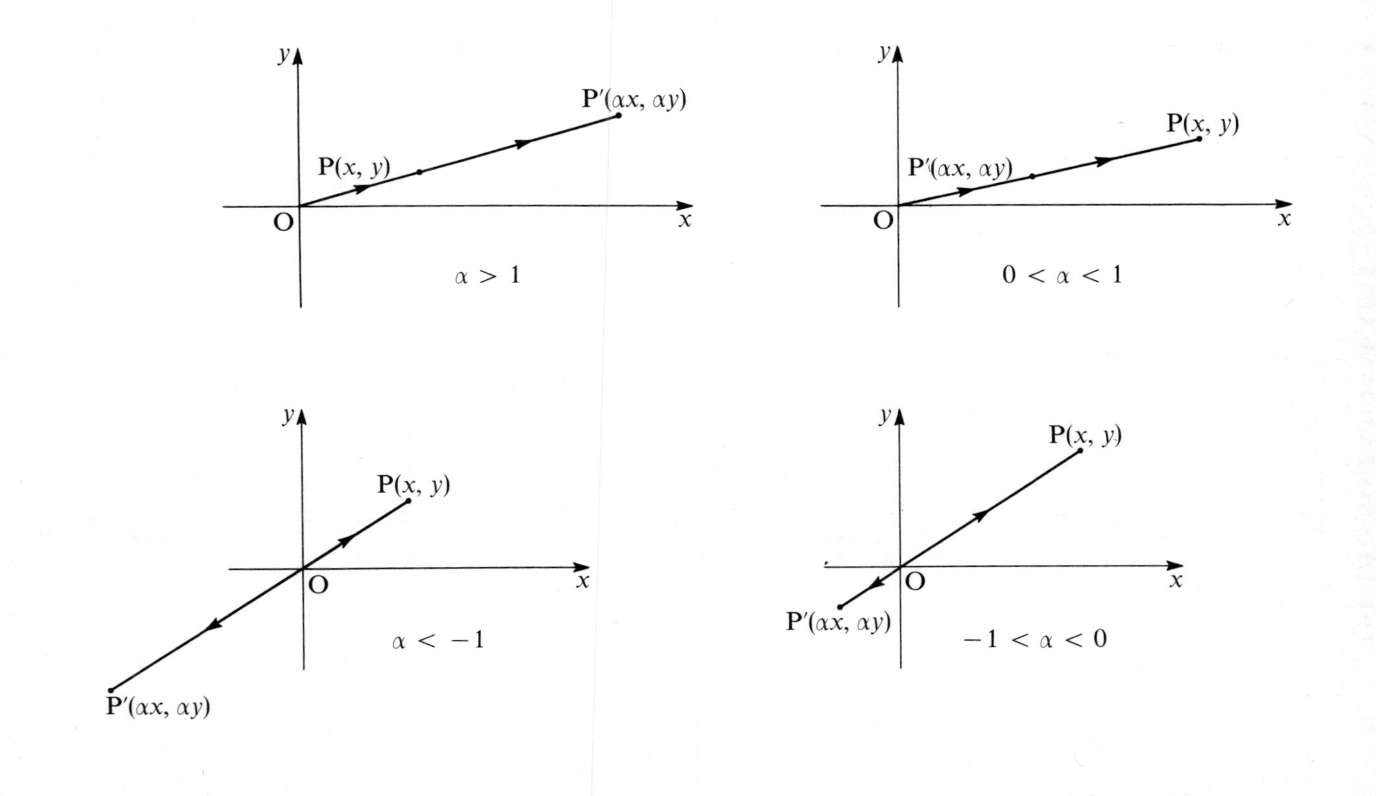

y
P′(αx, αy)
P(x, y)
O
x
α > 1
y
P(x, y)
P′(αx, αy)
O
x
0 < α < 1
y
P(x, y)
O
x
α < −1
P′(αx, αy)
y
P(x, y)
O
x
P′(αx, αy)
−1 < α < 0

$$\begin{aligned}
&= \alpha((x+x', y+y')) \\
&= (\alpha(x+x'), \alpha(y+y')) \\
&= (\alpha x+\alpha x', \alpha y+\alpha y') \quad \text{(by distributivity in } \mathbf{R}\text{)} \\
&= (\alpha x, \alpha y)+(\alpha x', \alpha y') \\
&= \alpha(x, y)+\alpha(x', y') \\
&= \alpha\mathbf{u}+\alpha\mathbf{v}.
\end{aligned}$$

Also

$$\begin{aligned}
(\alpha+\beta)\mathbf{u} &= (\alpha+\beta)(x, y) \\
&= ((\alpha+\beta)x, (\alpha+\beta)y) \\
&= (\alpha x+\beta x, \alpha y+\beta y) \quad \text{(by distributivity in } \mathbf{R}\text{)} \\
&= (\alpha x, \alpha y)+(\beta x, \beta y) \\
&= \alpha(x, y)+\beta(x, y) \\
&= \alpha\mathbf{u}+\beta\mathbf{u}.
\end{aligned}$$

Finally

$$\begin{aligned}
(\alpha\beta)\mathbf{u} &= (\alpha\beta)(x, y) \\
&= ((\alpha\beta)x, (\alpha\beta)y) \\
&= (\alpha(\beta x), \alpha(\beta y)) \quad \text{(by associativity in } \mathbf{R}\text{)} \\
&= \alpha(\beta x, \beta y) \\
&= \alpha(\beta(x, y)) \\
&= \alpha(\beta\mathbf{u}),
\end{aligned}$$

and

$$\begin{aligned}
1\mathbf{u} &= 1(x, y) \\
&= (1x, 1y) \\
&= (x, y) \quad \text{(by definition of 1 in } \mathbf{R}\text{)} \\
&= \mathbf{u}.
\end{aligned}$$

The algebraic rules in this theorem are perhaps rather different from those we have met before; in the first place we are combining *different* things, namely real numbers and pairs of real numbers. Nevertheless in form they bear strong similarities to some of our previous rules and are probably best remembered by recalling these similarities. For convenience we draw them together as a definition of what we now call a *real two-dimensional vector space.* To emphasize that this space is constructed from pairs of real numbers, we denote it by $\mathbf{R}^2$. In this context the real numbers in $\mathbf{R}$ are referred to as *scalars* so as to distinguish them from the vectors of $\mathbf{R}^2$.

Summing up, we have the following.

Definition. The real two-dimensional vector space $\mathbf{R}^2$ consists of ordered pairs (x, y) of real numbers, together with the rules for vector addition:

$$(x_1, y_1)+(x_2, y_2) = (x_1+x_2, y_1+y_2),$$

for all vectors (x_1, y_1) and (x_2, y_2) in $\mathbf{R}^2$, and for scalar multiplication:

$$\alpha(x, y) = (\alpha x, \alpha y),$$

for all vectors (x, y) in $\mathbf{R}^2$ and all scalars α in $\mathbf{R}$.

With these rules we have the following properties of $\mathbf{R}^2$:

(i) $\mathbf{u}+\mathbf{v} = \mathbf{v}+\mathbf{u}$ for all vectors $\mathbf{u}$, $\mathbf{v}$ in $\mathbf{R}^2$;

(ii) $(\mathbf{u}+\mathbf{v})+\mathbf{w} = \mathbf{u}+(\mathbf{v}+\mathbf{w})$ for all vectors $\mathbf{u}$, $\mathbf{v}$, $\mathbf{w}$ in $\mathbf{R}^2$;

(iii) there is a unique vector $\mathbf{0}$ in $\mathbf{R}^2$ such that

$$\mathbf{0}+\mathbf{u} = \mathbf{u} = \mathbf{u}+\mathbf{0}$$

for all $\mathbf{u}$ in $\mathbf{R}^2$;

(iv) for all vectors $\mathbf{u}$ in $\mathbf{R}^2$ there exists a unique vector $(-\mathbf{u})$ in $\mathbf{R}^2$ such that

$$\mathbf{u}+(-\mathbf{u}) = \mathbf{0} = (-\mathbf{u})+\mathbf{u};$$

(v) for all vectors $\mathbf{u}$, $\mathbf{v}$ in $\mathbf{R}^2$ and all scalars α, β in $\mathbf{R}$,

$$\alpha(\mathbf{u}+\mathbf{v}) = \alpha\mathbf{u}+\alpha\mathbf{v},$$

$$(\alpha+\beta)\mathbf{u} = \alpha\mathbf{u}+\beta\mathbf{u};$$

(vi) for all vectors $\mathbf{u}$ in $\mathbf{R}^2$ and all scalars α, β in $\mathbf{R}$,

$$(\alpha\beta)\mathbf{u} = \alpha(\beta\mathbf{u}),$$

(vii) for all vectors $\mathbf{u}$ in $\mathbf{R}^2$,

$$1\mathbf{u} = \mathbf{u}.$$

Finally we add a further set of properties of $\mathbf{R}^2$, based on the algebraic equivalent of the length of a displacement. We define what we call the *norm* of a vector in $\mathbf{R}^2$ as follows.

Definition. The norm of a vector $\mathbf{u} = (x, y)$, in $\mathbf{R}^2$, is denoted by $\|\mathbf{u}\|$ and given by

$$\|\mathbf{u}\| = \|(x, y)\| = \sqrt{(x^2+y^2)}.$$

The norm is thus a non-negative real number, which clearly, in geometrical illustrations of $\mathbf{R}^2$, will be interpreted as a length of some appropriate displacement. On the other hand, algebraically, we need to know how the norm fits with our rules for $\mathbf{R}^2$. Thus we ask how $\|u\|$ and $\|v\|$ are related to $\|u+v\|$, and how α and $\|u\|$ are related to $\|\alpha u\|$.

Theorem. *For any scalar α, with absolute value $|\alpha|$, and any vectors $\mathbf{u}_1 = (x_1, y_1)$, $\mathbf{u}_2 = (x_2, y_2)$, $\mathbf{u} = (x, y)$ in $\mathbf{R}^2$,*

(i) $\|\mathbf{u}_1+\mathbf{u}_2\| \leqslant \|\mathbf{u}_1\|+\|\mathbf{u}_2\|$,

(ii) $\|\mathbf{u}_1-\mathbf{u}_2\| \geqslant \big|\,\|\mathbf{u}_1\|-\|\mathbf{u}_2\|\,\big|$

(iii) $\|\alpha\mathbf{u}\| = |\alpha|\,\|\mathbf{u}\|$

Proof.

(i) By definition $\mathbf{u}_1+\mathbf{u}_2 = (x_1+x_2, y_1+y_2)$, so

$$\begin{aligned}\|\mathbf{u}_1+\mathbf{u}_2\|^2 &= (x_1+x_2)^2+(y_1+y_2)^2\\ &= (x_1^2+y_1^2)+2(x_1x_2+y_1y_2)+(x_2^2+y_2^2).\end{aligned}$$

But $(x_1^2+y_1^2)(x_2^2+y_2^2)-(x_1x_2+y_1y_2)^2 = (y_1x_2-x_1y_2)^2$,

so $(x_1x_2+y_1y_2) \leqslant \sqrt{((x_1^2+y_1^2)(x_2^2+y_2^2))} = \sqrt{(\|\mathbf{u}_1\| \, \|\mathbf{u}_2\|)}$,

and $\|\mathbf{u}_1+\mathbf{u}_2\|^2 \leqslant \|\mathbf{u}_1^2\| + 2\sqrt{(\|\mathbf{u}_1\| \, \|\mathbf{u}_2\|)} + \|\mathbf{u}_2^2\|$

$$= (\|\mathbf{u}_1\| + \|\mathbf{u}_2\|)^2.$$

Taking square roots, we have

$$\|\mathbf{u}_1+\mathbf{u}_2\| \leqslant \|\mathbf{u}_1\| + \|\mathbf{u}_2\|,$$

as required.

(ii) We deduce this result from (i), first replacing $\mathbf{u}_1$ by $\mathbf{u}_1-\mathbf{u}_2$, so that

$$\|\mathbf{u}_1-\mathbf{u}_2+\mathbf{u}_2\| = \|\mathbf{u}_1-\mathbf{u}_2\| + \|\mathbf{u}_2\|$$

i.e. $$\|\mathbf{u}_1\| - \|\mathbf{u}_2\| \leqslant \|\mathbf{u}_1-\mathbf{u}_2\|,$$

and then replacing $\mathbf{u}_2$ by $\mathbf{u}_2-\mathbf{u}_1$ in (i), so that

$$\|\mathbf{u}_1+\mathbf{u}_2-\mathbf{u}_1\| \leqslant \|\mathbf{u}_1\| + \|\mathbf{u}_2-\mathbf{u}_1\|,$$

i.e. $$\|\mathbf{u}_2\| - \|\mathbf{u}_1\| \leqslant \|\mathbf{u}_2-\mathbf{u}_1\|.$$

Since $\|\mathbf{u}_1-\mathbf{u}_2\| = \|\mathbf{u}_2-\mathbf{u}_1\|$, we thus have the two inequalities:

$$\|\mathbf{u}_1-\mathbf{u}_2\| \geqslant \|\mathbf{u}_1\| - \|\mathbf{u}_2\|$$

$$\|\mathbf{u}_1-\mathbf{u}_2\| \geqslant \|\mathbf{u}_2\| - \|\mathbf{u}_1\|.$$

We take the positive right-hand side to obtain

$$\|\mathbf{u}_1-\mathbf{u}_2\| \geqslant \big|\, \|\mathbf{u}_1\| - \|\mathbf{u}_2\| \,\big|.$$

(iii) We have

$$\begin{aligned} \|\alpha\mathbf{u}\| &= \|\alpha(x, y)\| \\ &= \|(\alpha x, \alpha y)\| \\ &= \sqrt{((\alpha x)^2+(\alpha y)^2)} \\ &= \sqrt{(\alpha^2(x^2+y^2))} \\ &= |\alpha|\sqrt{(x^2+y^2)} \\ &= |\alpha| \, \|\mathbf{u}\|. \end{aligned}$$

Reference back to the geometrical notion of length from which we obtained our idea of a norm, reveals that the first and last parts of this theorem, at least, are trivial when considered geometrically. Thus, in the case of the first part of the theorem, if (x_1, y_1), (x_2, y_2) are given vectors, we may illustrate them in the usual way by means of displacements in the plane, using the familiar parallelogram rule to illustrate their sum (x_1+x_2, y_1+y_2).

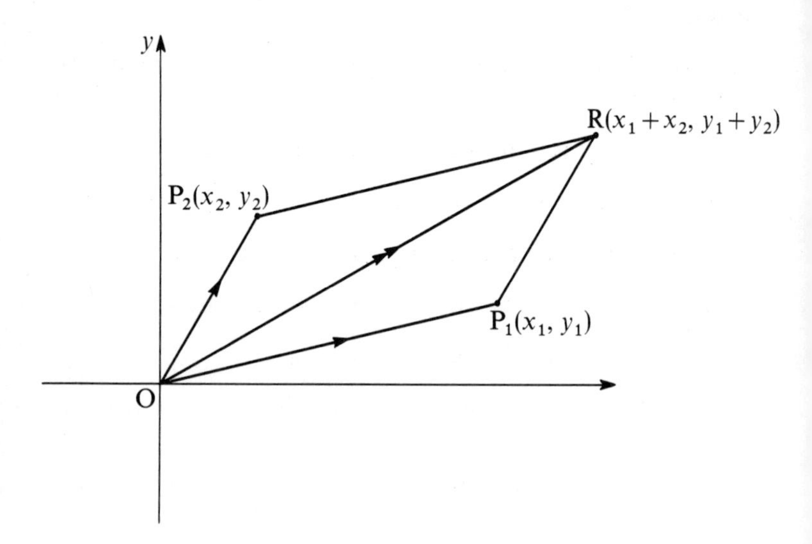

Now since in Euclidean geometry we prove that two sides of a triangle are together greater in length than the third side, we have OP_1 ($=\|(x_1, y_1)\|$ in length) and $P_1R = OP_2$ ($=\|(x_2, y_2)\|$ in length) together greater in length than OR ($=\|(x_1+x_2, y_1+y_2\|$ in length). We therefore have

$$\|(x_1, y_1)\| + \|(x_2, y_2)\| \geqslant \|(x_1+x_2, y_1+y_2)\|.$$

Similarly (but even more simply), as far as the last part of the theorem is concerned, if we multiply a vector by a scalar we change the length of the corresponding displacement by a factor equal to the absolute value of the scalar, i.e. $\|\alpha(x, y)\| = |\alpha| \, \|(x, y)\|$.

Worked examples 2.1

1 Compute

$$(1, 2)-(0, 1)+\tfrac{1}{2}(2, 4)-2(1, \tfrac{1}{2})$$

in $\mathbf{R}^2$. Illustrate your answer.

$$\begin{aligned}(1, 2)-(0, 1)+\tfrac{1}{2}(2, 4)-2(1, \tfrac{1}{2}) &= (1, 2)+(0, -1)+(1, 2)+(-2, -1)\\ &= (0, 2).\end{aligned}$$

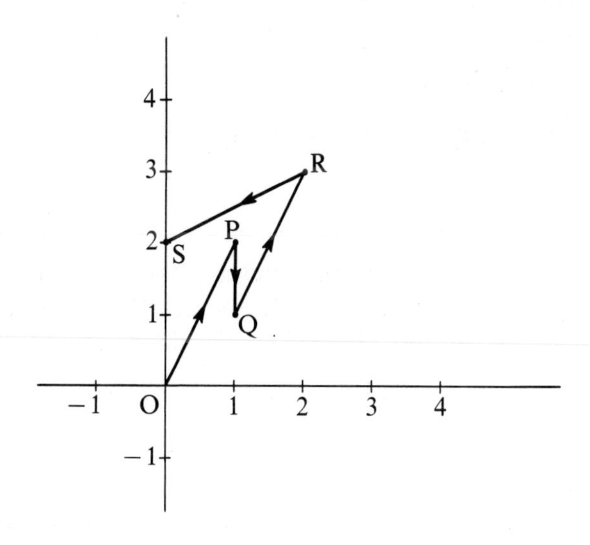

$\overrightarrow{OP}$ corresponds to $(1, 2)$

$\overrightarrow{PQ}$ corresponds to $-(0, 1)$

$\overrightarrow{QR}$ corresponds to $\frac{1}{2}(2, 4)$

$\overrightarrow{RS}$ corresponds to $-2(1, \frac{1}{2})$

$\overrightarrow{OS}$ corresponds to $(0, 2)$

$\overrightarrow{OS} = \overrightarrow{OP}+\overrightarrow{PQ}+\overrightarrow{QR}+\overrightarrow{RS}$

2 Displacements $\overrightarrow{OP}$, $\overrightarrow{OQ}$ and $\overrightarrow{OR}$ in the real co-ordinate plane correspond to vectors $(1, 2)$, $(-1, 3)$ and $(\frac{1}{2}, \frac{1}{3})$ in $\mathbf{R}^2$. What are the vectors corresponding to the displacements $\overrightarrow{PQ}$, $\overrightarrow{RQ}$ and $\overrightarrow{PR}$?

$\overrightarrow{PQ}$ is the displacement corresponding to the vector $-(1, 2)+(-1, 3) = (-2, 1)$, since the displacement $-\overrightarrow{OP}$ when added to the displacement $\overrightarrow{OQ}$ yields the displacement $\overrightarrow{PQ}$ (we can write $\quad = \overrightarrow{PO}$ $-\overrightarrow{OP}+\overrightarrow{OQ}$).

$\overrightarrow{RQ}$ is the displacement corresponding to

$$-(\tfrac{1}{2}, \tfrac{1}{3})+(-1, 3) = (1-1\tfrac{1}{2}, 2\tfrac{2}{3}),$$

since $\overrightarrow{RQ} = \overrightarrow{RO}+\overrightarrow{OQ} = -\overrightarrow{OR}+\overrightarrow{OQ}$.

$\overrightarrow{PR}$ corresponds to $-(1, 2)+(\frac{1}{2}, \frac{1}{3}) = (-\frac{1}{2}, -1\frac{2}{3})$.

3 If $\overrightarrow{OP}$ and $\overrightarrow{OQ}$ are displacements in the real co-ordinate plane corresponding to vectors $\mathbf{u}$, $\mathbf{v}$ in $\mathbf{R}^2$, describe R in relation to P and Q if $\overrightarrow{OR}$ is the displacement corresponding to the vector $\dfrac{r}{r+s}\mathbf{u}+\dfrac{s}{r+s}\mathbf{v}$, where $r+s \neq 0$.

If P has co-ordinates (x, y), and $Q(x', y')$ then R has co-ordinates $\left(\dfrac{rx+sx'}{r+s}, \dfrac{ry+sy'}{r+s}\right)$, since $\overrightarrow{OR}$ corresponds to the vector

$$\frac{r}{r+s}(x, y)+\frac{s}{r+s}(x', y') = \left(\frac{rx+sx'}{r+s}, \frac{ry+sy'}{r+s}\right).$$

It follows from elementary co-ordinate geometry that R divides PQ in the ratio $s:r$.

4 Prove that for all $\mathbf{u}$ in $\mathbf{R}^2$, $0\mathbf{u} = \mathbf{0}$ and $(-1)\mathbf{u} = -\mathbf{u}$.

Suppose $\mathbf{u} = (x, y)$. Then

$$\begin{aligned} 0\mathbf{u} &= 0(x, y) \\ &= (0x, 0y) \\ &= (0, 0) \\ &= \mathbf{0}. \end{aligned}$$

Again

$$\begin{aligned} (-1)\mathbf{u} &= (-1)(x, y) \\ &= ((-1)x, (-1)y) \\ &= (-x, -y) \\ &= -\mathbf{u}. \end{aligned}$$

Exercises 2.1

1 Compute in $\mathbf{R}^2$
(i) $(3, \frac{1}{2}) - 5(1, 1) + \frac{1}{2}(10, 6) - (7, 1)$,
(ii) $\frac{1}{2}((2, 3) - (\sqrt{2}, \pi))$,
(iii) $(1, 0) + 3(0, 1) - (1, 3)$.

2 Illustrate the fact that in $\mathbf{R}^2$,
(i) $\alpha(\mathbf{u}+\mathbf{v}) = \alpha\mathbf{u} + \alpha\mathbf{v}$,
(ii) $(\alpha+\beta)\mathbf{u} = \alpha\mathbf{u} + \beta\mathbf{u}$.

3 If displacements $\overrightarrow{OP}$ and $\overrightarrow{OQ}$ in the real co-ordinate plane correspond to vectors $\mathbf{u}$ and $\mathbf{v}$ in $\mathbf{R}^2$, construct the point R in case $\overrightarrow{OR}$ is the displacement corresponding to $3\mathbf{u} - \frac{1}{2}\mathbf{v}$.

4 The displacement $\overrightarrow{OP}$ corresponds in the real co-ordinate plane to a vector $\mathbf{u}$ of length 2 in $\mathbf{R}^2$. If P moves

anticlockwise around the circle centre O, radius 2, and $\overrightarrow{OQ}$ is the displacement corresponding to a vector $\mathbf{v}$, describe in geometric terms the set of all points Q,
(i) if $\mathbf{v} = \mathbf{u}+(1, 0)$,
(ii) if $\mathbf{v} = 2\mathbf{u}$,
(iii) if $\mathbf{v} = -\mathbf{u}$.

5 Prove that if $\mathbf{u}$ is a vector in $\mathbf{R}^2$, and α is a scalar in $\mathbf{R}$, then
(i) $(-\alpha)\mathbf{u} = -(\alpha\mathbf{u})$,
(ii) $\alpha \neq 0$, $\mathbf{u} \neq \mathbf{0}$ implies $\alpha\mathbf{u} \neq \mathbf{0}$.

6 Prove that if $\mathbf{u}$ is a fixed non-zero vector in $\mathbf{R}^2$, then the set of vectors $\alpha\mathbf{u}$, where α is any scalar in $\mathbf{R}$, is closed under vector addition and scalar multiplication in $\mathbf{R}^2$.
Is the same true for vectors $\alpha\mathbf{u}+(1, 0)$?

7 Prove that any vector in $\mathbf{R}^2$ can be written in the form $\alpha(3, 1)+\beta(1, 2)$, for suitable choice of scalars α, β.
Is the same true if we replace (3, 1) and (1, 2) by (1, 4) and (3, 12) respectively?

2.2 Linear dependence and independence in $\mathbf{R}^2$

Given vectors $\mathbf{u}_1$, $\mathbf{u}_2$, ..., $\mathbf{u}_m$ in $\mathbf{R}^2$ and scalars, i.e. real numbers, α_1, α_2, ..., α_m in $\mathbf{R}$, we can form the vector

$$\alpha_1\mathbf{u}_1+\alpha_2\mathbf{u}_2+\ldots+\alpha_m\mathbf{u}_m.$$

Indeed, this is the most complicated algebraic expression we can form using the operations of addition and scalar multiplication of vectors in $\mathbf{R}^2$.

If, given $\mathbf{u}_1$, $\mathbf{u}_2$, $\cdots$, $\mathbf{u}_m$ in $\mathbf{R}^2$, we can find scalars α_1, α_2, ..., α_m in $\mathbf{R}$, *not all zero*, such that

$$\alpha_1\mathbf{u}_1+\alpha_2\mathbf{u}_2+\ldots+\alpha_m\mathbf{u}_m = \mathbf{0},$$

we say that the vectors $\mathbf{u}_1$, $\mathbf{u}_2$, ..., $\mathbf{u}_m$ are *linearly dependent*. If on the contrary, again given $\mathbf{u}_1$, $\mathbf{u}_2$, ..., $\mathbf{u}_m$ in $\mathbf{R}^2$, the equation

$$\alpha_1\mathbf{u}_1+\alpha_2\mathbf{u}_2+\ldots+\alpha_m\mathbf{u}_m = \mathbf{0}$$

implies $\alpha_1 = \alpha_2 = \ldots = \alpha_m = 0$, we say the vectors $\mathbf{u}_1, \mathbf{u}_2, \ldots, \mathbf{u}_m$ are *linearly independent.*

If $\mathbf{u}_1, \mathbf{u}_2, \ldots, \mathbf{u}_m$ are linearly dependent we say they form a *linearly dependent set of vectors,* $\{\mathbf{u}_1, \mathbf{u}_2, \ldots, \mathbf{u}_m\}$. Of course the set $\{0\}$, consisting of the zero vector alone must then be considered an acceptable, but possibly uninteresting linearly dependent set of vectors. Similarly we can have *linearly independent sets of vectors* $\{\mathbf{u}_1, \mathbf{u}_2, \ldots, \mathbf{u}_m\}$, and a set $\{\mathbf{u}\}$, where $\mathbf{u}$ is any non-zero vector in $\mathbf{R}^2$, is a linearly independent set.

It is easy to give examples of linearly dependent and linearly independent vectors in $\mathbf{R}^2$. Thus (2, 4) and (3, 6) are linearly dependent since

$$\begin{aligned} 3(2, 4)+(-2)(3, 6) &= (6, 12)+(-6, -12) \\ &= (6-6, 12-12) \\ &= (0, 0). \end{aligned}$$

Or again

$$\begin{aligned} (2, 4)+(-\tfrac{2}{3})(3, 6) &= (2, 4)+(-2, -4) \\ &= (2-2, 4-4) \\ &= (0, 0). \end{aligned}$$

Similarly, (1, 2), (1, 0) and (0, 1) are linearly dependent, since

$$\begin{aligned} (1, 2)+(-1)(1, 0)+(-2)(0, 1) &= (1, 2)+(-1, 0)+(0, -2) \\ &= (1-1, 2-2) \\ &= (0, 0). \end{aligned}$$

On the other hand (1, 0) and (0, 1) are linearly independent, for if

$$\alpha(1, 0)+\beta(0, 1) = (0, 0)$$

for some real numbers α, β, then

$$(\alpha, 0)+(0, \beta) = (0, 0),$$

i.e.

$$(\alpha, \beta) = (0, 0),$$

whence $\alpha = \beta = 0$.

The notions of linear dependence and independence are also usefully illustrated geometrically. Consider two non-zero vectors (x, y) and (x', y'), and their illustration in the real co-ordinate plane by means of displacements $\overrightarrow{OP}$ and $\overrightarrow{OQ}$, where P has co-ordinates (x, y) and Q has co-ordinates (x', y'). The vectors (x, y) and (x', y') are dependent if and only if we have scalars α, β, not both zero, such that

$$\alpha(x, y)+\beta(x', y') = (0, 0).$$

But then
$$(\alpha x+\beta x', \alpha y+\beta y') = (0, 0),$$

i.e.
$$\alpha x+\beta x' = 0,$$

and
$$\alpha y+\beta y' = 0.$$

In this case, therefore, $\dfrac{x}{x'} = \dfrac{y}{y'}$ or equivalently $\dfrac{x}{y} = \dfrac{x'}{y'}$, and our knowledge of simple co-ordinate geometry tells us that the points O, P and Q are collinear. Appropriate illustrations for dependent vectors (x, y), (x', y') in $\mathbf{R}^2$ can therefore take the following forms.

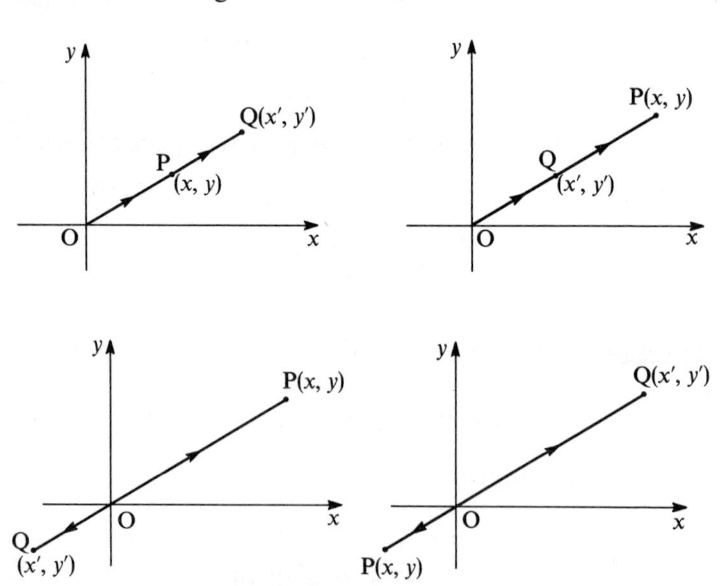

Dependent vectors

On the other hand, if two vectors are independent, they are not dependent and the movements which represent them cannot be collinear. The appropriate illustration for independent vectors (x, y) and (x', y') in $\mathbf{R}^2$ is thus of the following form.

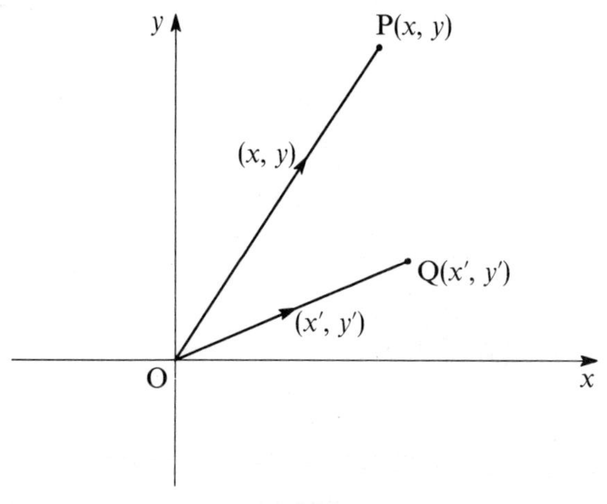

Independent vectors

It follows from the above discussion that if we have three vectors in $\mathbf{R}^2$, no two of which are linearly dependent, we have an associated illustration of the following form.

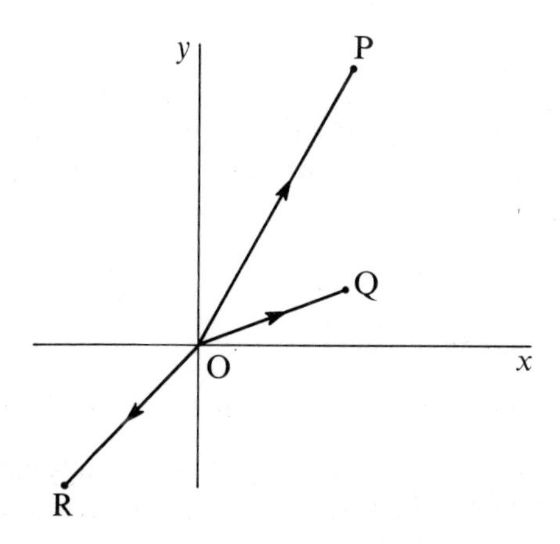

Using results from elementary plane geometry we can complete such an illustration so as to obtain a parallelogram OP′RQ′ as follows.

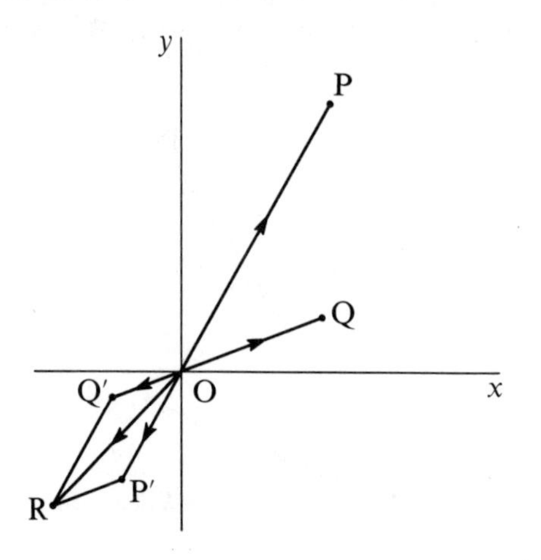

In co-ordinate geometrical terms, then, if P, Q and R have co-ordinates (x, y), (x', y') and (x'', y'') respectively, then P′ has co-ordinates $(\alpha x, \alpha y)$ for some real number α, and Q′ has co-ordinates $(\beta x', \beta y')$ for some real number β. But since OP′RQ′ is a parallelogram, this implies

$$x'' = \alpha x + \beta x'$$

and

$$y'' = \alpha y + \beta y'.$$

Thus our illustration suggests the truth of the statement that if (x, y), (x', y'), (x'', y'') are three vectors in $\mathbf{R}^2$, no two of which are linearly dependent, then there exist real numbers α, β such that

$$(x'', y'') = (\alpha x, \alpha y) + (\beta x', \beta y'),$$

or equivalently, such that

$$(-1)(x'', y'') + \alpha(x, y) + \beta(x', y') = (0, 0).$$

In other words three vectors in $\mathbf{R}^2$ would appear, of necessity, to be dependent, however much we try to keep them independent.

From these intuitive geometrical arguments we are thus led to formulate and prove the following result.

Theorem. *Three or more non-zero vectors in $\mathbf{R}^2$ are linearly dependent.*

This theorem, suggested to us by plane geometrical arguments, must of course now be proved algebraically; i.e. we must deduce it by logical argument from our definition of the vector space $\mathbf{R}^2$.

Before giving the proof we prove two lemmas and a corollary which will help us in proving the theorem.

Lemma 1. *Any set of vectors in $\mathbf{R}^2$ which contains a dependent set of vectors is itself a dependent set.*

Proof. Suppose $\{\mathbf{u}_1, \mathbf{u}_2, \ldots, \mathbf{u}_n\}$ is a set of vectors containing the dependent set $\{\mathbf{u}_{i_1}, \mathbf{u}_{i_2}, \ldots, \mathbf{u}_{i_m}\}$. By hypothesis there exist scalars $\alpha_{i_1}, \alpha_{i_2}, \ldots, \alpha_{i_m}$, not all zero, such that

$$\alpha_{i_1}\mathbf{u}_{i_1} + \alpha_{i_2}\mathbf{u}_{i_2} + \ldots + \alpha_{i_m}\mathbf{u}_{i_m} = \mathbf{0}.$$

Thus setting $\alpha_j = 0$, $j \neq i_1, i_2, \ldots, i_m$, we have a linear dependence

$$\alpha_1\mathbf{u}_1 + \alpha_2\mathbf{u}_2 + \ldots + \alpha_n\mathbf{u}_n = 0,$$

which proves the result.

Corollary. *Any set of vectors in $\mathbf{R}^2$ which contains the zero vector is a dependent set.*

Proof. We observed above that the set $\{\mathbf{0}\}$, consisting of the zero vector alone, is a dependent set, so the corollary follows immediately from the lemma.

Lemma 2. *Any two vectors of the form* $(0, y_1)$, $(0, y_2)$ in $\mathbf{R}^2$ *are dependent.*

Proof. If both the vectors are non-zero, then $y_1 \neq 0$ and $y_2 \neq 0$, and we have the linear dependence

$$(-y_2)(0, y_1)+(y_1)(0, y_2) = (0, 0).$$

If one of them is zero, the result follows immediately from the corollary to Lemma 1.

We now prove the theorem. Because of Lemma 1, we prove only that any *three* vectors in $\mathbf{R}^2$ are dependent. Because of the corollary to Lemma 1 we assume all the vectors are non-zero, otherwise the result follows immediately.

Proof of theorem. Let (x_1, y_1), (x_2, y_2), (x_3, y_3) be any three non-zero vectors in $\mathbf{R}^2$. Either all of x_1, x_2, x_3 are zero, or one at least is non-zero. If $x_1 = x_2 = x_3 = 0$, our vectors are of the form $(0, y_1)$, $(0, y_2)$, $(0, y_3)$. But then, by Lemma 2, any pair of them is dependent, so by Lemma 1 the whole set of vectors is dependent and the truth of the theorem is established in this case. On the other hand if one of x_1, x_2, x_3 is non-zero, say x_1, it has an inverse $x_1^{-1} = 1/x_1$ in $\mathbf{R}$. Consider then the two vectors

$$(x_2, y_2)-\frac{x_2}{x_1}(x_1, y_1) = \left(0, y_2-\frac{x_2y_1}{x_1}\right),$$

and
$$(x_3, y_3)-\frac{x_3}{x_1}(x_1, y_1) = \left(0, y_3-\frac{x_3y_1}{x_1}\right).$$

By Lemma 2 these two vectors are dependent, i.e. there exist real numbers α, β, not both zero, such that

$$\alpha\left((x_2, y_2)-\frac{x_2}{x_1}(x_1, y_1)\right)+\beta\left((x_3, y_3)-\frac{x_3}{x_1}(x_1, y_1)\right) = (0, 0).$$

It follows immediately that

$$\left(-\frac{\alpha x_2}{x_1}-\frac{\beta x_3}{x_1}\right)(x_1, y_1)+\alpha(x_2, y_2)+\beta(x_3, y_3) = (0, 0),$$

but this is a linear dependence between (x_1, y_1), (x_2, y_2) and (x_3, y_3), which completes our proof.

Worked examples 2.2

1 Exhibit a linear dependence between the vectors (1, 2) and (2, 4), in $\mathbf{R}^2$.

There are an infinity of possible answers, among which are

$$(-2)(1, 2)+(2, 4) = (0, 0),$$

and
$$\tfrac{1}{2}(1, 2)+(-\tfrac{1}{4})(2, 4) = (0, 0).$$

2 Prove or disprove the linear independence of the following sets of vectors in $\mathbf{R}^2$:

(i) $\{(\frac{1}{2}, \frac{1}{3}), (5, 2)\}$,
(ii) $\{(4, -1), (\sqrt{2}, 7), (0, 0)\}$.

(i) If $\alpha(\frac{1}{2}, \frac{1}{3})+\beta(5, 2) = (0, 0)$

then
$$\tfrac{1}{2}\alpha+5\beta = 0,$$

and
$$\tfrac{1}{3}\alpha+2\beta = 0.$$

But these equations imply $\alpha = \beta = 0$, so the given set of vectors is linearly independent.

(ii) Any set of three vectors in $\mathbf{R}^2$ must be dependent, but, more than that, any set of vectors containing the zero vector (0, 0) must be a dependent set. Thus in this case we have a dependence

$$0(4, -1)+0(\sqrt{2}, 7)+\alpha(0, 0) = (0, 0),$$

whatever non-zero scalar α is chosen.

3 Given any two vectors (x_1, y_1) and (x_2, y_2) in $\mathbf{R}^2$, construct geometrically the vector

$$(x_2, y_2)-\frac{x_2}{x_1}(x_1, y_1)$$

introduced on page 88 in the proof of the theorem in this section.

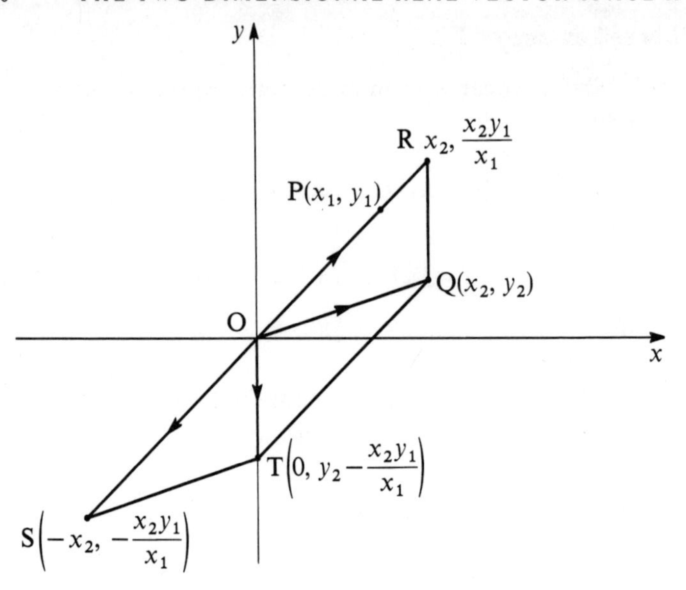

Given P and Q, we construct R by drawing QR parallel to the y-axis, meeting OP in R. The co-ordinates of R are $\left(x_2, \frac{x_2 y_1}{x_1}\right)$. We then take the point S such that OR = OS, and complete the parallelogram OSTQ. By construction T lies on the y-axis, and $\overrightarrow{OT}$ represents $(x_2, y_2) - \left(x_2, \frac{x_2 y_1}{y_1}\right) = (x_2, y_2) - \frac{x_2}{x_1}(x_1, y_1)$.

Exercises 2.2

1 Which of the following sets of vectors in $\mathbf{R}^2$ are linearly independent? If they are linearly independent prove them to be so; if they are dependent exhibit a dependence.

(i) $\{(1, 2)\}$,
(ii) $\{(\sqrt{2}, 1), (-2, 3)\}$,
(iii) $\{(\frac{1}{2}, \frac{1}{3}), (3, 2)\}$,
(iv) $\left\{\left(-\frac{1}{\sqrt{3}}, \frac{2}{\sqrt{3}}\right), (15, -30)\right\}$.

Illustrate each of your answers by drawing movements in the real co-ordinate plane representing displacements corresponding to the given vectors.

2 Illustrate by geometrical constructions the dependence of the various vectors introduced on page 88, in the proof of the theorem in this section, in case
(i) $(x_1, y_1) = (1, 0), (x_2, y_2) = (0, 3), (x_3, y_3) = (-1, -2)$.
(ii) $(x_1, y_1) = (2, 3), (x_2, y_2) = (4, -2), (x_3, y_3) = (-6, 1)$.

2.3 Bases and co-ordinates in $\mathbf{R}^2$

Our work so far with dependent and independent vectors has surely made it clear that although any three vectors in $\mathbf{R}^2$ must be dependent, a pair of vectors taken at random may or may not be dependent. Thus already, in Lemma 2, and before that, we have given examples of pairs of dependent and independent vectors, and of course there are many such pairs.

In particular we have noted that $(0, 1)$ and $(1, 0)$ are independent. We may also note that given any vector (a, b) in $\mathbf{R}^2$, the dependence between the vectors $(1, 0)$, $(0, 1)$ and (a, b) is easy to write down, since

$$(a, b) = a(1, 0)+b(0, 1),$$

so that

$$\begin{aligned} a(1, 0)+b(0, 1)+(-1)(a, b) &= (a, b)+(-a, -b) \\ &= (0, 0). \end{aligned}$$

Because we can write the dependence in the form

$$(a, b) = a(1, 0)+b(0, 1),$$

we say (a, b) is *dependent* on $(1, 0)$ and $(0, 1)$.

A pair of vectors in $\mathbf{R}^2$, with the properties that they are independent and such that any vector in $\mathbf{R}^2$ is dependent on them, is called a *basis* for $\mathbf{R}^2$. We have exhibited the particular basis $\{(1, 0), (0, 1)\}$, for $\mathbf{R}^2$, because it so obviously has these required properties. It is called the *standard* basis for $\mathbf{R}^2$. More generally, the theorem of section 2.2 implies that *any* pair of independent vectors in $\mathbf{R}^2$ can be taken as a basis for $\mathbf{R}^2$. For suppose (a_1, b_1) and (a_2, b_2) are independent and (a, b) is any vector in $\mathbf{R}^2$. By our theorem, there exist scalars α, β, γ, not all zero, such that

$$\alpha(a_1, b_1)+\beta(a_2, b_2)+\gamma(a, b) = (0, 0).$$

But since (a_1, b_1) and (a_2, b_2) are not dependent, we may assume $\gamma \neq 0$ (otherwise taking $\gamma = 0$ yields a forbidden dependence). Thus we have an inverse for γ, in $\mathbf{R}$, and we have

$$(a, b) = \left(-\frac{\alpha}{\gamma}\right)(a_1, b_1)+\left(-\frac{\beta}{\gamma}\right)(a_2, b_2).$$

Of course, we have done no more than use the independence of the pair (a_1, b_1), (a_2, b_2) to rewrite the dependence given to us by our theorem.

Now let us suppose we are given a basis $\{\mathbf{u}_1, \mathbf{u}_2\}$ for $\mathbf{R}^2$, and a vector $\mathbf{v}$ in $\mathbf{R}^2$ such that

$$\mathbf{v} = x\mathbf{u}_1+y\mathbf{u}_2,$$

for some real numbers x, y. Could we also have

$$\mathbf{v} = x'\mathbf{u}_1+y'\mathbf{u}_2,$$

where $x \neq x'$, $y \neq y'$? No, since the two equations imply

$$(x-x')\mathbf{u}_1+(y-y')\mathbf{u}_2 = \mathbf{0},$$

and since $\mathbf{u}_1$ and $\mathbf{u}_2$ are independent, this implies $x-x' = 0$ and $y-y' = 0$, i.e. $x = x'$ and $y = y'$. Thus given a basis $\{\mathbf{u}_1, \mathbf{u}_2\}$ for $\mathbf{R}^2$ we have associated with any vector $\mathbf{v}$ in $\mathbf{R}^2$ a unique pair of real numbers x, y such that

$$\mathbf{v} = x\mathbf{u}_1+y\mathbf{u}_2.$$

We refer to x and y as the *co-ordinates of the vector* $\mathbf{v}$ *with respect to the basis* $\{\mathbf{u}_1, \mathbf{u}_2\}$. Notice it is most important that we give the basis in a particular order; the co-ordinates of $\mathbf{v}$ with respect to the basis $\{\mathbf{u}_1, \mathbf{u}_2\}$, *taken in that order*, will not in general be the same as the co-ordinates with respect to the basis $\{\mathbf{u}_2, \mathbf{u}_1\}$ *taken in that order*.

The reader should also note that the co-ordinates of a vector (a, b) with respect to the standard basis $\{(1, 0), (0, 1)\}$, taken in that order, are a and b respectively. For this basis, and for this basis only, the co-ordinates of any vector (a, b) in $\mathbf{R}^2$ are the same as the real numbers a and b which define the vector itself, and incidentally the same as the co-ordinates (a, b) of the point we use in representing the vector geometrically in the real co-ordinate plane.

It is sometimes convenient to work with a basis other than the standard basis. When this is to be done, it is usual to specify the chosen basis once for all and then to work with the co-ordinates with respect to that basis. In such a case then, each vector in $\mathbf{R}^2$ is referred to by its co-ordinates in the new basis instead of in its original form. This is entirely justifiable since, as we have seen, a vector uniquely determines its co-ordinates in any given basis, and conversely, given a basis, a vector is itself uniquely determined by its co-ordinates with respect to that basis.

As far as notation is concerned, once a new basis has been specified, we use the same notation as before, so that if

$$\mathbf{v} = x\mathbf{u}_1 + y\mathbf{u}_2,$$

with respect to a basis $\{\mathbf{u}_1, \mathbf{u}_2\}$, we write (x, y) for the vector $\mathbf{v}$ and suppress the vectors $\mathbf{u}_1$ and $\mathbf{u}_2$. No confusion arises provided it is clear which basis we are working in. Thus if $\mathbf{v}$ is the vector (a, b) with respect to the standard basis, we do not write $(a, b) = (x, y)$. If such a formula was required, we would probably 'label' the brackets to say which basis we were in, e.g. by using $(\ ,\ )_{\mathrm{S}}$ for the standard basis and $(\ ,\ )_{\mathrm{B}}$ for a basis B. We could then write $(a, b)_{\mathrm{S}} = (x, y)_{\mathrm{B}}$, if necessary.

The algebraic rules for manipulation of the co-ordinates (x, y) in a given basis other than the standard one are the same as for vectors in $\mathbf{R}^2$, for if with respect to a basis $\{\mathbf{u}_1, \mathbf{u}_2\}$ we have

$$\mathbf{v}_1 = x_1\mathbf{u}_1 + y_1\mathbf{u}_2,$$

and

$$\mathbf{v}_2 = x_2\mathbf{u}_1 + y_2\mathbf{u}_2,$$

then
$$\mathbf{v}_1+\mathbf{v}_2 = x_1\mathbf{u}_1+y_1\mathbf{u}_2+x_2\mathbf{u}_1+y_2\mathbf{u}_2$$
$$= (x_1+x_2)\mathbf{u}_1+(y_1+y_2)u_2,$$

so that in the basis $\mathbf{u}_1$, $\mathbf{u}_2$ we can write

$$(x_1, y_1)+(x_2, y_2) = (x_1+x_2, y_1+y_2),$$

just as we do when we are working with the standard basis. Similarly for scalar multiplication

$$\alpha(x\mathbf{u}_1+y\mathbf{u}_2) = (\alpha x)\mathbf{u}_1+(\alpha y)\mathbf{u}_2,$$

so that
$$\alpha(x, y) = (\alpha x, \alpha y).$$

Thus, apart from possible occasional convenience (as in picking the best axes in setting about a co-ordinate geometrical problem), there is nothing to choose between one basis and another. Of course $\mathbf{R}^2$ comes to us equipped with its standard basis, but there are no algebraic procedural changes to be made if we pick another basis and work with co-ordinates with respect to this basis rather than with the standard one.

Worked examples 2.3

1 Find the co-ordinates of the vector (1, 0) in $\mathbf{R}^2$, with respect to the (ordered) basis $\{(1, 1), (-1, 1)\}$.

Illustrate your answer.

We require x, y such that

$$(1, 0) = x(1, 1)+y(-1, 1).$$

Since $x(1, 1)+y(-1, 1) = (x-y, x+y)$,

we require to solve the equations

$$1 = x-y$$
$$0 = x+y.$$

The required co-ordinates are therefore $x = \frac{1}{2}$, $y = -\frac{1}{2}$.

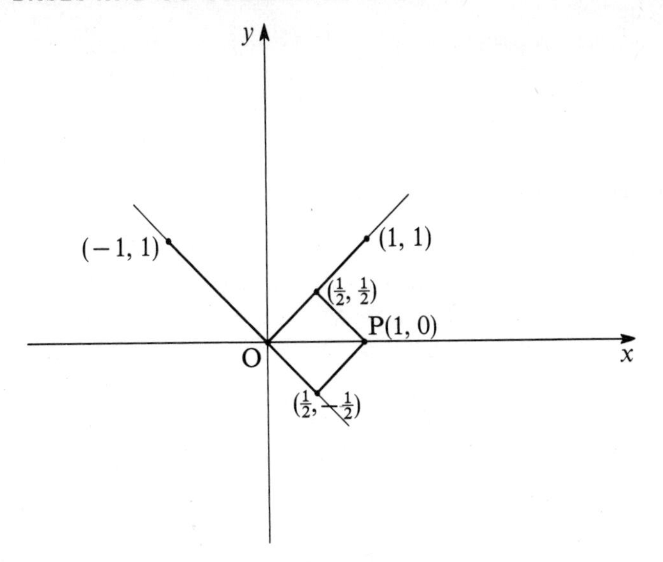

2 Pick a basis for $\mathbf{R}^2$ from the vectors $(0, 1)$, $(-2, 3)$, $(2, -4)$, and $(0, -1)$. Will any pair of these vectors *not* form a basis?

We require two independent vectors. Starting with any one of the given vectors, we select a second to form an independent set. So we take $(0, 1)$, and check that

$$\alpha(0, 1)+\beta(-2, 3) = (0, 0)$$

implies

$$(-2\beta, \alpha+3\beta) = (0, 0),$$

i.e. $\alpha = \beta = 0$, so that $(-2, 3)$ is independent of $(0, 1)$ and together they will form a basis.

The only pair of vectors in the given set which is dependent is $(0, 1)$, $(0, -1)$, for which

$$(0, 1)+(0, -1) = (0, 0).$$

Exercises 2.3

In each of the following exercises find the co-ordinates of the given vector with respect to the given (ordered) basis. Illustrate your answers.

1 $(1, 0)$ with respect to $\{(0, 1), (1, 0)\}$.
2 (a, b) with respect to $\{(1, 1), (-1, 1)\}$.
3 $(2, -4)$ with respect to $\{(0, 1), (-2, 3)\}$.

Miscellaneous exercises 2

1 If $\{\mathbf{u}, \mathbf{v}, \mathbf{w}\}$ is a dependent set of vectors in $\mathbf{R}^2$, must $\mathbf{u}$ be dependent on $\mathbf{v}$ and $\mathbf{w}$? Can $\{\mathbf{u}, \mathbf{v}, \mathbf{w}\}$ be a dependent set of vectors in $\mathbf{R}^2$ such that no one of the vectors is dependent on the other two?

2 Find a basis for $\mathbf{R}^2$ with respect to which the co-ordinates of any vector (a, b) are α, β where

$$\alpha = 2a-b,$$

and

$$\beta = a+b.$$

3 Two bases, $\{\mathbf{u}_1, \mathbf{v}_1\}$ and $\{\mathbf{u}_2, \mathbf{v}_2\}$ are given in $\mathbf{R}^2$. With respect to the first basis, a vector $\mathbf{w}$ has co-ordinates (a_1, b_1):

$$\mathbf{w} = a_1\mathbf{u}_1 + b_1\mathbf{v}_1.$$

With respect to the second basis, $\mathbf{w}$ has co-ordinates (a_2, b_2):

$$\mathbf{w} = a_2\mathbf{u}_2 + b_2\mathbf{v}_2.$$

If

$$\mathbf{u}_2 = p\mathbf{u}_1 + q\mathbf{v}_1,$$

and

$$\mathbf{v}_2 = r\mathbf{u}_1 + s\mathbf{v}_1,$$

express a_1 and b_1 in terms of a_2 and b_2.

4 Prove that real numbers of the form $a+b\sqrt{2}$, where a and b are any rational numbers, together with the operations of real addition and multiplication, form a *vector space over the field of rational numbers*, i.e. they have properties (i), $\cdots$, (vii) on pages 77, 78, in relation to the (scalar) field of rational numbers.

CHAPTER THREE

Multiplicative structure in $\mathbf{R}^2$

3.1 Expected properties of multiplication in $\mathbf{R}^2$

So far, in studying $\mathbf{R}^2$, we have found that vectors can be added together and multiplied by scalars. Our definition of addition has been seen to ensure reasonable properties, already more or less familiar to us from our previous knowledge of addition in fields of numbers. Our scalar multiplication has led us into possibly newer territory, since although its properties may perhaps seem familiar, it involves a multiplication of different kinds of mathematical objects, namely real numbers and ordered pairs of real numbers, rather than objects of the same kind.

In this chapter we are concerned to ask, and answer, the question 'why should we not multiply two vectors in $\mathbf{R}^2$ together so as to obtain a product which is a vector in $\mathbf{R}^2$?': we can certainly multiply numbers in a field of numbers; why not vectors in the vector space $\mathbf{R}^2$?

Of course, the question is rather too broad unless we insist that if an algebraic process is to be called 'multiplication', then it must have some, if not all, of the usual algebraic properties of a multiplication in a field of numbers. In fact, in the case of $\mathbf{R}^2$, not only must a multiplication have 'reasonable' properties in itself, but it must also 'fit' with our vector addition and scalar multiplication, just as multiplication and addition of numbers in a field fit together.

What then ought we to expect of the multiplication we seek? If we are to be able to perform all our usual algebraic processes we must insist on commutativity, associativity, a multiplicative identity, and multiplicative inverses (for non-zero vectors at least). In symbols, given any vectors $\mathbf{u}$, $\mathbf{v}$, $\mathbf{w}$ in $\mathbf{R}^2$, we expect

$$\mathbf{uv} = \mathbf{vu};$$

$$(\mathbf{uv})\mathbf{w} = \mathbf{u}(\mathbf{vw});$$

$$\text{a vector } \mathbf{1} \neq \mathbf{0} \text{ such that } \mathbf{1u} = \mathbf{u} = \mathbf{u1};$$

$$\text{if } \mathbf{u} \neq \mathbf{0}, \text{ a vector } \mathbf{u}^{-1} \text{ such that } \mathbf{uu}^{-1} = \mathbf{1} = \mathbf{u}^{-1}\mathbf{u}.$$

Already in making use of a 'non-zero' condition, both for the identity and for inverses in our multiplication, we have a link with vector addition, for it is in the properties of addition that the zero vector is to be found originally. Other links must also be expected, e.g. we must expect $\mathbf{u0} = \mathbf{0} = \mathbf{0u}$, for all vectors $\mathbf{u}$. More generally we must also expect distributivity of multiplication over addition, i.e., symbolically, we expect

$$\mathbf{u}(\mathbf{v}+\mathbf{w}) = \mathbf{uv}+\mathbf{uw},$$

and

$$(\mathbf{u}+\mathbf{v})\mathbf{w} = \mathbf{uw}+\mathbf{vw},$$

for any vectors $\mathbf{u}$, $\mathbf{v}$, $\mathbf{w}$ in $\mathbf{R}^2$.

All these conditions would lead us in the first place to look for a situation typified for us already by our properties of fields, where addition and multiplication and their properties are familiar to us. Since, as we have already noted, in the vector space $\mathbf{R}^2$ we have scalar multiplication, if all our structures are to 'fit' together, then we must also expect links between vector multiplication and scalar multiplication.

What then should we expect of a product of scalar multiples such as $(\alpha\mathbf{u})(\beta\mathbf{v})$? If α and β are positive integers the answer is clear, for our definition of scalar multiplication in $\mathbf{R}^2$ implies that if m is an integer

$$\begin{aligned} m\mathbf{u} &= (1+1+\ldots+1)\mathbf{u} \quad (m \text{ summands}) \\ &= 1\mathbf{u}+1\mathbf{u}+\ldots+1\mathbf{u} \\ &= \mathbf{u}+\mathbf{u}+\ldots+\mathbf{u}, \end{aligned}$$

so that for example

$$\begin{aligned} 2\mathbf{u} &= \mathbf{u}+\mathbf{u}, \\ 3\mathbf{v} &= \mathbf{v}+\mathbf{v}+\mathbf{v}, \end{aligned}$$

and

$$\begin{aligned} (2\mathbf{u})(3\mathbf{v}) &= (\mathbf{u}+\mathbf{u})(\mathbf{v}+\mathbf{v}+\mathbf{v}) \\ &= \mathbf{uv}+\mathbf{uv}+\ldots+\mathbf{uv}, \\ &= 6\mathbf{uv}, \end{aligned}$$

assuming our multiplication to be distributive over addition. Thus for integer scalars α, β we expect

$$(\alpha\mathbf{u})(\beta\mathbf{v} = (\alpha\beta)(\mathbf{uv}).$$

But if integer scalar multiples must satisfy such a condition we should surely expect other scalar multiples to behave similarly. Thus we expect that for all scalars α, β and all vectors $\mathbf{u}$, $\mathbf{v}$, our multiplication will fit with the scalar multiplication in $\mathbf{R}^2$, in the sense that $(\alpha\mathbf{u})(\beta\mathbf{v}) = (\alpha\beta)(\mathbf{uv})$.

Let us now collect together these various properties: we expect a 'reasonable' multiplication of vectors in $\mathbf{R}^2$ to associate with any vectors $\mathbf{u}$, $\mathbf{v}$ in $\mathbf{R}^2$ a unique vector, written $\mathbf{uv}$, called the product of $\mathbf{u}$ and $\mathbf{v}$, and such that

(i) for all $\mathbf{u}$, $\mathbf{v}$ in $\mathbf{R}^2$, $\mathbf{uv} = \mathbf{vu}$;

(ii) for all $\mathbf{u}$, $\mathbf{v}$, $\mathbf{w}$ in $\mathbf{R}^2$, $(\mathbf{uv})\mathbf{w} = \mathbf{u}(\mathbf{vw})$;

(iii) there shall exist in $\mathbf{R}^2$ a unique vector $\mathbf{1} \neq \mathbf{0}$, such that for all $\mathbf{u}$ in $\mathbf{R}^2$, $\mathbf{u1} = \mathbf{u} = \mathbf{1u}$;

(iv) for any non-zero $\mathbf{u}$ in $\mathbf{R}^2$, there shall exist a unique vector $\mathbf{u}^{-1}$ in $\mathbf{R}^2$ such that $\mathbf{u}^{-1}\mathbf{u} = \mathbf{1} = \mathbf{uu}^{-1}$;

(v) for all $\mathbf{u}$, $\mathbf{v}$, $\mathbf{w}$ in $\mathbf{R}^2$,

$$\begin{aligned} \mathbf{u}(\mathbf{v}+\mathbf{w}) &= \mathbf{uv}+\mathbf{uw}, \\ (\mathbf{u}+\mathbf{v})\mathbf{w} &= \mathbf{uw}+\mathbf{vw}; \end{aligned}$$

(vi) for all scalars α, β, and all $\mathbf{u}$, $\mathbf{v}$ in $\mathbf{R}^2$,

$$(\alpha\mathbf{u})(\beta\mathbf{v}) = (\alpha\beta)(\mathbf{uv}).$$

As usual, commutativity and associativity will be taken to imply that rearrangements of the order of any number of vectors in a product and of the order of multiplication of pairs of vectors in the product do not affect the product.

Worked examples 3.1

1 In our expected property (v) of multiplication in $\mathbf{R}^2$ we have assumed distributivity of multiplication over addition, i.e. that for all vectors $\mathbf{u}$, $\mathbf{v}$, $\mathbf{w}$ in $\mathbf{R}^2$, we have

$$\mathbf{u}(\mathbf{v}+\mathbf{w}) = \mathbf{uv}+\mathbf{uw}$$

and

$$(\mathbf{u}+\mathbf{v})\mathbf{w} = \mathbf{uw}+\mathbf{vw}.$$

Given our other assumed properties of multiplication, show that it is not reasonable to ask for distributivity of addition over multiplication, i.e. reversing the rules of addition and multiplication in the above equations, to ask that for all $\mathbf{u}$, $\mathbf{v}$, $\mathbf{w}$ in $\mathbf{R}^2$,

$$\mathbf{u}+(\mathbf{vw}) = (\mathbf{u}+\mathbf{v})(\mathbf{u}+\mathbf{w})$$

and

$$\mathbf{uv}+\mathbf{w} = (\mathbf{u}+\mathbf{w})(\mathbf{v}+\mathbf{w}).$$

Taking $\mathbf{u} = \mathbf{v} = \mathbf{1}$ and $\mathbf{w} = \mathbf{0}$,

$$\begin{aligned}\mathbf{u}+(\mathbf{vw}) &= \mathbf{1}+(\mathbf{1})(\mathbf{0})\\ &= \mathbf{1}+\mathbf{0}\\ &= \mathbf{1},\end{aligned}$$

by our assumed properties of $\mathbf{1}$. But

$$\begin{aligned}(\mathbf{u}+\mathbf{v})(\mathbf{u}+\mathbf{w}) &= (\mathbf{1}+\mathbf{1})(\mathbf{1}+\mathbf{0})\\ &= (\mathbf{1}+\mathbf{1})\mathbf{1}\\ &= \mathbf{1}+\mathbf{1}.\end{aligned}$$

Thus, if we assume

$$\mathbf{u}+(\mathbf{vw}) = (\mathbf{u}+\mathbf{v})(\mathbf{u}+\mathbf{w}),$$

we must assume

$$\mathbf{1} = \mathbf{1}+\mathbf{1},$$

i.e.

$$\mathbf{1} = \mathbf{0},$$

which contradicts the non-zero assumption about $\mathbf{1}$.

A similar contradiction can be obtained in the case of the second equation by taking $\mathbf{u} = \mathbf{0}$ and $\mathbf{v} = \mathbf{w} = \mathbf{1}$.

2 Deduce from our expected properties (i)–(vi), of multiplication in $\mathbf{R}^2$, that $\mathbf{0u} = \mathbf{0}$, for all $\mathbf{u}$ in $\mathbf{R}^2$, if such a multiplication exists.

The arguments which were used for the integers and for the real numbers can be used again here. Thus

$$(\mathbf{0}+\mathbf{u})\mathbf{u} = \mathbf{uu} \qquad (\text{since } \mathbf{0}+\mathbf{u} = \mathbf{u}),$$

so

$$\mathbf{0u}+\mathbf{uu} = \mathbf{uu} \qquad (\text{by distributivity})$$

Adding $-(\mathbf{uu})$ to both sides of this equation and using associativity of vector addition, we deduce $\mathbf{0u} = \mathbf{0}$ as required.

Exercises 3.1

1 Prove that if there is a multiplication in $\mathbf{R}^2$ with the expected properties (i)–(vi) above, then
(a) $(-\mathbf{u})(\mathbf{v}) = -(\mathbf{uv})$,
(b) $(-\mathbf{u})^{-1} = -(\mathbf{u}^{-1})$,
(c) $-(\alpha\mathbf{u}) = (-\alpha)\mathbf{u}$,
for all vectors $\mathbf{u}$, $\mathbf{v}$ and scalars α.

2 What consequences, if any, would you consider unreasonable if multiplications in $\mathbf{R}^2$ were defined by setting
(i) $\mathbf{uv} = \mathbf{u}+\mathbf{v}$,

(ii) $\mathbf{uv} = \|\mathbf{u}\| \mathbf{v}$,
(iii) $(x_1, y_1)(x_2, y_2) = (x_1x_2, y_1y_2)$,
for all vectors $\mathbf{u}$, $\mathbf{v}$, (x_1, y_1) and (x_2, y_2)?

3.2 Multiplication and linear dependence in $\mathbf{R}^2$

In this section we consider the further implications of there being defined on $\mathbf{R}^2$ a multiplication which has the properties we have listed above. In particular we shall consider how such a multiplication is affected by the theory of linear dependence which we developed for $\mathbf{R}^2$ in the last chapter. Eventually we shall find that any multiplication will have to take the form

$$(a, b)(c, d) = (ac - bd, bc + ad),$$

i.e. will in effect have to be our usual complex multiplication.

Later, in a converse direction, we shall find that such a formula for complex multiplication does indeed define a means of multiplying vectors in $\mathbf{R}^2$ which has all the properties we require. The necessary checks to show this are simple and straightforward (but essential, for until they are made we do not know that there is even one satisfactory multiplication!). Now, however, for the time being we assume there is given a multiplication having the required properties, but without a specific formula.

We begin, then, by considering the vector $\mathbf{1}$, which exists by property (iii) on page 99. Suppose $\mathbf{1} = (x, y)$. Since $\mathbf{1} \neq \mathbf{0}$, at least one of the real numbers x, y must be non-zero. If $x \neq 0$, we can easily construct a vector $\mathbf{u}$ in $\mathbf{R}^2$ independent of the vector $\mathbf{1}$. Thus $\mathbf{u} = (x, y+1)$ will suffice, for if there exist real numbers α, β such that

$$\alpha(x, y) + \beta(x, y+1) = \mathbf{0},$$

then
$$(\alpha x + \beta x, \alpha y + \beta y + \beta) = (0, 0),$$

i.e.
$$(\alpha + \beta)x = 0 \quad \text{and} \quad (\alpha + \beta)y + \beta = 0.$$

But $x \neq 0$, so $\alpha + \beta = 0$, whence $\beta = 0$ and in consequence $\alpha = 0$. Similarly if $y \neq 0$, (x, y) and $(x+1, y)$ are independent.

Given the vector $\mathbf{1} \neq \mathbf{0}$, in $\mathbf{R}^2$, we can therefore assume there exists a vector $\mathbf{u}$, in $\mathbf{R}^2$, independent of $\mathbf{1}$. But then according to the results of Chapter two, the pair of vectors $\{\mathbf{1}, \mathbf{u}\}$ forms a basis for $\mathbf{R}^2$. Thus, if we multiply $\mathbf{u}$ by itself, the vector $\mathbf{uu}$ must be dependent on the vectors $\mathbf{1}$ and $\mathbf{u}$, i.e. there exist unique scalars α, β such that

$$\mathbf{uu} = \alpha\mathbf{1} + \beta\mathbf{u}.$$

Now $-(\alpha\mathbf{1}) = (-\alpha)\mathbf{1}$ and $-(\beta\mathbf{u}) = (-\beta)\mathbf{u}$, so that if we write $-\alpha = c$, $-\beta = b$ and $\mathbf{u}^2 = \mathbf{uu}$, we can rearrange this dependence in the form of a 'quadratic' in $\mathbf{u}$:

$$\mathbf{u}^2 + b\mathbf{u} + c\mathbf{1} = \mathbf{0}.$$

If the properties we expect our multiplication to have are 'reasonable', we would surely expect to be able to perform standard algebraic processes, such as 'completing the square', on such a quadratic. Taking perhaps more care than usual, so as to see how we use the properties of our multiplication, let us try to do exactly that.

First, using distributivity,

$$\begin{aligned}\left(\mathbf{u}+\frac{b}{2}\mathbf{1}\right)\left(\mathbf{u}+\frac{b}{2}\mathbf{1}\right) &= \left(\mathbf{u}+\frac{b}{2}\mathbf{1}\right)\mathbf{u}+\left(\mathbf{u}+\frac{b}{2}\mathbf{1}\right)\left(\frac{b}{2}\mathbf{1}\right)\\ &= \mathbf{uu}+\left(\frac{b}{2}\mathbf{1}\right)\mathbf{u}+\mathbf{u}\left(\frac{b}{2}\mathbf{1}\right)+\left(\frac{b}{2}\mathbf{1}\right)\left(\frac{b}{2}\mathbf{1}\right).\end{aligned}$$

Using commutativity this becomes

$$\mathbf{uu}+\left(\frac{b}{2}\mathbf{1}\right)\mathbf{u}+\left(\frac{b}{2}\mathbf{1}\right)\mathbf{u}+\left(\frac{b}{2}\mathbf{1}\right)\left(\frac{b}{2}\mathbf{1}\right),$$

which again using distributivity becomes

$$\mathbf{uu}+\left(\frac{b}{2}\mathbf{1}\right)(\mathbf{u}+\mathbf{u})+\left(\frac{b}{2}\mathbf{1}\right)\left(\frac{b}{2}\mathbf{1}\right) = \mathbf{uu}+\left(\frac{b}{2}\mathbf{1}\right)(2\mathbf{u})+\left(\frac{b}{2}\mathbf{1}\right)\left(\frac{b}{2}\mathbf{1}\right).$$

Using the properties $(\alpha\mathbf{u})(\beta\mathbf{v}) = (\alpha\beta)(\mathbf{uv})$, $\mathbf{1u} = \mathbf{u}$, and $(\mathbf{1})(\mathbf{1}) = \mathbf{1}$, this last expression can be written

$$\mathbf{uu} + b\mathbf{u} + \frac{b^2}{4}\mathbf{1}.$$

Thus

$$\mathbf{u}^2+b\mathbf{u}+c\mathbf{1} = \left(\mathbf{u}+\frac{b}{2}\mathbf{1}\right)\left(\mathbf{u}+\frac{b}{2}\mathbf{1}\right)-\left(\frac{b^2}{4}\mathbf{1}\right)+c\mathbf{1},$$

$$= \left(\mathbf{u}+\frac{b}{2}\mathbf{1}\right)\left(\mathbf{u}+\frac{b}{2}\mathbf{1}\right)+\left(c-\frac{b^2}{4}\right)\mathbf{1},$$

since $-\left(\frac{b^2}{4}\mathbf{1}\right) = \left(-\frac{b^2}{4}\right)\mathbf{1}$. As we expected, using the properties of our multiplication, we have indeed been able to 'complete the square' associated with our 'quadratic' dependence.

Can we proceed further along this traditional algebraic road and in some sense attempt to 'solve' our quadratic dependence? To simplify notation, let us write Δ for the real number $c-\frac{b^2}{4}$. We then have

$$\left(\mathbf{u}+\frac{b}{2}\mathbf{1}\right)\left(\mathbf{u}+\frac{b}{2}\mathbf{1}\right)+\Delta\mathbf{1} = \mathbf{0}.$$

Since the real field $\mathbf{R}$ is ordered, either $\Delta \leqslant 0$ or $\Delta > 0$. Suppose $\Delta \leqslant 0$. Then $-\Delta \geqslant 0$ and, as we have already noted on page 49, this implies that the square root $\sqrt{(-\Delta)}$ exists in $\mathbf{R}$. In this case then, again using our assumed properties† for multiplication we can 'factorize' our dependence as a product:

$$\left(\left(\mathbf{u}+\frac{b}{2}\mathbf{1}\right)+\sqrt{(-\Delta)}\mathbf{1}\right)\left(\left(\mathbf{u}+\frac{b}{2}\mathbf{1}\right)-\sqrt{(-\Delta)}\mathbf{1}\right) = \mathbf{0}.$$

But if this equation is true, and both factors on the left are non-zero, we have here an equation of the form

$$\mathbf{wv} = \mathbf{0}, \quad \mathbf{w} \neq \mathbf{0}, \quad \mathbf{v} \neq \mathbf{0}.$$

But then $\mathbf{w}$ has an inverse, being non-zero, and since we assume our multiplication will be associative, we have

† Which specific properties are used here?

$$\begin{aligned}(\mathbf{w}^{-1})\mathbf{w}\mathbf{v} &= (\mathbf{w}^{-1}\mathbf{w})\mathbf{v}\\ &= \mathbf{1}\mathbf{v}\\ &= \mathbf{v},\end{aligned}$$

so that $\mathbf{w}\mathbf{v} = \mathbf{0}$ implies

$$\begin{aligned}\mathbf{v} &= \mathbf{w}^{-1}\mathbf{0}\\ &= \mathbf{0}\end{aligned}$$

(by Worked example 2, p. 101)

which contradicts our assumption $\mathbf{v} \neq \mathbf{0}$, so that if $\mathbf{w}\mathbf{v} = 0$ then either $\mathbf{w} = 0$ or $\mathbf{v} = 0$.

Thus if $\Delta \leqslant 0$, an attempt to factorize our dependence yields either

$$\mathbf{u}+\frac{b}{2}\mathbf{1}+\sqrt{(-\Delta)}\mathbf{1} = \mathbf{0} \quad \text{or} \quad \mathbf{u}+\frac{b}{2}\mathbf{1}-\sqrt{(-\Delta)}\mathbf{1} = \mathbf{0}.$$

But neither of these equations is possible, since the vectors $\mathbf{u}$ and $\mathbf{1}$ are linearly independent. Thus our assumption $\Delta \leqslant 0$ leads to a contradiction.

We can therefore only presume $\Delta > 0$. In this case we have in $\mathbf{R}$ the square root $\sqrt{\Delta}$, and again using our assumed properties[†] of multiplication, we can write the dependence

$$\left(\mathbf{u}+\frac{b}{2}\mathbf{1}\right)\left(\mathbf{u}+\frac{b}{2}\mathbf{1}\right)+\Delta\mathbf{1} = \mathbf{0},$$

in the form

$$\left(\frac{1}{\sqrt{\Delta}}\mathbf{u}+\frac{b}{2\sqrt{\Delta}}\mathbf{1}\right)^2+\mathbf{1} = \mathbf{0}.$$

Now the vectors $\mathbf{1}$ and $\left(\frac{1}{\sqrt{\Delta}}\mathbf{u}+\frac{b}{2\sqrt{\Delta}}\mathbf{1}\right)$ are independent, for if

$$\alpha\mathbf{1}+\beta\left(\frac{1}{\sqrt{\Delta}}\mathbf{u}+\frac{b}{2\sqrt{\Delta}}\mathbf{1}\right) = \mathbf{0},$$

[†] Again, which specific properties are used here?

we have

$$\left(\alpha+\frac{\beta b}{2\sqrt{\Delta}}\right)\mathbf{1}+\frac{\beta}{\sqrt{\Delta}}\mathbf{u} = \mathbf{0}$$

and the independence of **1** and **u** then implies, as required, that $\alpha = \beta = 0$. Let us therefore take the vectors **1** and $\left(\frac{1}{\sqrt{\Delta}}\mathbf{u}+\frac{b}{2\sqrt{\Delta}}\mathbf{1}\right)$ as a basis for $\mathbf{R}^2$, in place of **1** and **u**. We write **i** for the second vector in this basis, so that

$$\mathbf{i} = \frac{1}{\sqrt{\Delta}}\mathbf{u}+\frac{b}{2\sqrt{\Delta}}\mathbf{1},$$

and we have $\mathbf{i}^2+\mathbf{1} = \mathbf{0}$.

Thus our assumption that a multiplication exists in $\mathbf{R}^2$, having the properties set out in the last section, leads us to the conclusion that in such a case there will be a basis for $\mathbf{R}^2$ consisting of vectors **1** and **i**, where **1** is the identity for our multiplication, and **i** is such that its square is $-\mathbf{1}$. We shall thus be able to write any vector in $\mathbf{R}^2$ in the form

$$\alpha\mathbf{1}+\beta\mathbf{i},$$

for some scalars α, β. Moreover our formula for multiplication will follow, from our properties, to be given in general by

$$\begin{aligned}(\alpha\mathbf{1}+\beta\mathbf{i})(\gamma\mathbf{1}+\delta\mathbf{i}) &= (\alpha\mathbf{1}+\beta\mathbf{i})(\gamma\mathbf{1})+(\alpha\mathbf{1}+\beta\mathbf{i})(\delta\mathbf{i}) \\ &= (\alpha\gamma)\mathbf{1}+(\beta\gamma)\mathbf{i}+(\alpha\delta)\mathbf{i}+(\beta\delta)(-\mathbf{1}) \\ &= (\alpha\gamma-\beta\delta)\mathbf{1}+(\beta\gamma+\alpha\delta)\mathbf{i}.\end{aligned}$$

It may be noted that in the final stages of our argument, we could equally well have taken a basis **1**, $-\left(\frac{1}{\sqrt{\Delta}}\mathbf{u}+\frac{b}{2\sqrt{\Delta}}\mathbf{1}\right)$. In this case, writing

$$\mathbf{i} = -\left(\frac{1}{\sqrt{\Delta}}\mathbf{u}+\frac{b}{2\sqrt{\Delta}}\mathbf{1}\right)$$

we find ourselves in exactly the same position as before, i.e. we have a basis $\{\mathbf{1}, \mathbf{i}\}$ for $\mathbf{R}^2$, with the property $\mathbf{i}^2+\mathbf{1} = \mathbf{0}$, or equivalently $\mathbf{i}^2 = -\mathbf{1}$. Thus, if this alternative choice of basis is taken, once again our formula for multiplication of vectors in $\mathbf{R}^2$ can be written

$$(\alpha\mathbf{1}+\beta\mathbf{i})(\gamma\mathbf{1}+\delta\mathbf{i}) = (\alpha\gamma-\beta\delta)\mathbf{1}+(\beta\gamma+\alpha\delta)\mathbf{i}.$$

Thus whichever of the bases, $\mathbf{1}$, $\left(\frac{1}{\sqrt{\Delta}}\mathbf{u}+\frac{b}{2\sqrt{\Delta}}\mathbf{1}\right)$ or $\mathbf{1}$, $-\frac{1}{\sqrt{\Delta}}\mathbf{u}+\frac{b}{2\sqrt{\Delta}}\mathbf{1}$, we choose to work with, *if we express our vectors in co-ordinate form with respect to the chosen basis, writing* $\alpha\mathbf{1}+\beta\mathbf{i}$ *in the form* (α, β), *we have*

$$(\alpha, \beta)(\gamma, \delta) = (\alpha\gamma-\beta\delta, \beta\gamma+\alpha\delta).$$

Worked examples 3.2

1 Deduce carefully from properties (i)–(vi) of a multiplication in $\mathbf{R}^2$ the correctness of our assertion, above, that

$$\left(\mathbf{u}+\frac{b}{2}\mathbf{1}\right)\left(\mathbf{u}+\frac{b}{2}\mathbf{1}\right)+\Delta\mathbf{1}$$
$$= \left(\left(\mathbf{u}+\frac{b}{2}\mathbf{1}\right)+\sqrt{(-\Delta)}\mathbf{1}\right)\left(\left(\mathbf{u}+\frac{b}{2}\mathbf{1}\right)-\sqrt{(-\Delta)}\mathbf{1}\right).$$

For ease of computation we change the notation and prove

$$\mathbf{v}^2+(-\alpha^2)\mathbf{1} = (\mathbf{v}+\alpha\mathbf{1})(\mathbf{v}-\alpha\mathbf{1}),$$

for any vector $\mathbf{v}$ and scalar α. By distributivity, and our rules for scalar multiplication, we have

$$(\mathbf{v}+\alpha\mathbf{1})(\mathbf{v}-\alpha\mathbf{1}) = \mathbf{v}(\mathbf{v}-\alpha\mathbf{1})+(\alpha\mathbf{1})(\mathbf{v}-\alpha\mathbf{1}),$$
$$= \mathbf{v}(\mathbf{v}+(-\alpha)\mathbf{1})+(\alpha\mathbf{1})(\mathbf{v}+(-\alpha)\mathbf{1}).$$

Since $\mathbf{1v} = \mathbf{v}$, again using distributivity, this last expression can be written

$$\mathbf{vv}+(\mathbf{1v})((-\alpha)\mathbf{1})+(\alpha\mathbf{1})(\mathbf{1v})+(\alpha\mathbf{1})((-\alpha)\mathbf{1}).$$

Since $(\alpha\mathbf{u})(\beta\mathbf{v}) = (\alpha\beta)\mathbf{uv}$, for all scalars α, β and vectors $\mathbf{u}$, $\mathbf{v}$, this in turn can be written

$$\mathbf{v}^2+(-\alpha)(\mathbf{v1})+\alpha(\mathbf{1v})+(-\alpha^2)((\mathbf{1})(\mathbf{1})).$$

But $\mathbf{1v} = \mathbf{v1} = \mathbf{v}$ for all vectors $\mathbf{v}$, so this is equal to

$$\mathbf{v}^2+(-\alpha)\mathbf{v}+\alpha\mathbf{v}+(-\alpha^2)\mathbf{1}.$$

Finally by our rules for scalar multiplication, $(-\alpha)\mathbf{v} = -(\alpha\mathbf{v})$, so that

$$(\mathbf{v}+\alpha\mathbf{1})(\mathbf{v}-\alpha\mathbf{1}) = \mathbf{v}^2+(-\alpha^2)\mathbf{1},$$

as required.

2 Compute, using the formula

$$(a, b)(c, d) = (ac-bd, bc+ad),$$

and the expected properties of our multiplication, the vector

$$((1, 2)+(1, -2))((0, 1)(3, 4))$$

$$\begin{aligned}((1, 2)+(1, -2))((0, 1)(3, 4)) &= ((2, 0))((-4, 3))\\ &= (-8, 6).\end{aligned}$$

Exercises 3.2

1 Which of the expected properties (i)–(vi) of a multiplication in $\mathbf{R}^2$ are required to justify our assertion, above, that

$$\left(\mathbf{u}+\frac{b}{2}\mathbf{1}\right)\left(\mathbf{u}+\frac{b}{2}\mathbf{1}\right)+\Delta\mathbf{1} = 0 \text{ implies} \left(\frac{1}{\sqrt{\Delta}}\mathbf{u}+\frac{b}{2\sqrt{\Delta}}\mathbf{1}\right)^2+\mathbf{1} = 0?$$

2 Compute, as in Worked example 2 immediately above, the vectors

(i) $(0, 1)(a, b)$

(ii) $(3(-1, 2)+(0, -1))((-2, 1)-(3, -2))$

(iii) $\left(-\frac{1}{2}, \frac{\sqrt{3}}{2}\right)^3$

(iv) $\left(-\frac{1}{2}, \frac{-\sqrt{3}}{2}\right)^3$

3.3 Construction of a multiplication in $\mathbf{R}^2$

We began this chapter by setting out in detail the algebraic properties we claimed we would expect of a 'reasonable' multiplication of vectors in $\mathbf{R}^2$. In the last section we deduced the algebraic form a multiplication has to take, *if it exists and has the given properties.*

Now we show there is such a multiplication. To be precise, we define a multiplication of vectors in $\mathbf{R}^2$ and check that it has all the properties we require of a multiplication if it is to be algebraically acceptable to us.

For completeness we begin by restating our list of required properties of a 'reasonable' multiplication, but now we make a new definition of them as follows.

Definition. The vector space $\mathbf{R}^2$ is said to carry the structure of a *division algebra over the field* $\mathbf{R}$, if there is defined in $\mathbf{R}^2$ a rule which associates with any two vectors $\mathbf{u}$, $\mathbf{v}$ in $\mathbf{R}^2$ a unique vector $\mathbf{uv}$ in $\mathbf{R}^2$, such that

(i) for all vectors $\mathbf{u}$, $\mathbf{v}$ in $\mathbf{R}^2$, $\mathbf{uv} = \mathbf{vu}$;
(ii) for all vectors $\mathbf{u}$, $\mathbf{v}$, $\mathbf{w}$ in $\mathbf{R}^2$, $(\mathbf{uv})\mathbf{w} = \mathbf{u}(\mathbf{vw})$;
(iii) there exists a unique non-zero vector $\mathbf{1}$ in $\mathbf{R}^2$, such that for all vectors $\mathbf{u}$ in $\mathbf{R}^2$, $\mathbf{u1} = \mathbf{u} = \mathbf{1u}$;
(iv) for any non-zero vector $\mathbf{u}$ in $\mathbf{R}^2$, there exists a unique vector $\mathbf{u}^{-1}$ in $\mathbf{R}^2$ such that $\mathbf{u}^{-1}\mathbf{u} = \mathbf{1} = \mathbf{uu}^{-1}$;
(v) for all vectors $\mathbf{u}$, $\mathbf{v}$, $\mathbf{w}$ in $\mathbf{R}^2$,

$$\mathbf{u}(\mathbf{v}+\mathbf{w}) = \mathbf{uv}+\mathbf{uw},$$

$$(\mathbf{u}+\mathbf{v})\mathbf{w} = \mathbf{uw}+\mathbf{vw};$$

(vi) for all real numbers α, β, and all $\mathbf{u}$, $\mathbf{v}$ in $\mathbf{R}^2$,

$$(\alpha\mathbf{u})(\beta\mathbf{v}) = (\alpha\beta)(\mathbf{uv}).$$

We now define a rule for multiplying vectors in $\mathbf{R}^2$.

Definition. *For all vectors* (a, b), (c, d) in $\mathbf{R}^2$, *we set*

$$(a, b)(c, d) = (ac - bd, bc + ad).$$

Clearly, our choice of this multiplicative rule is made in the light of our work in the previous section. There we proved that *if* a reasonable multiplication exists, then, after taking co-ordinates with respect to a suitable basis, it must be of this form. Here, we prove that there is a multiplication available, of the expected form and satisfying our required rules. The basis we are working with is of course the standard basis $\{(1, 0), (0, 1)\}$, with which $\mathbf{R}^2$ is always equipped. We repeat that we must prove an 'existence' result, for otherwise there may be no multiplication with the desired properties.

We formulate our result as a theorem.

Theorem. *The multiplication defined by setting*

$$(a, b)(c, d) = (ac - bd, bc + ad)$$

for all elements (a, b), (c, d) *in* $\mathbf{R}^2$, *gives* $\mathbf{R}^2$ *the structure of a division algebra over* $\mathbf{R}$.

Proof. We need to check each of conditions (i)–(vi) in the preceding definition of a division algebra over $\mathbf{R}$.

(i)
$$\begin{aligned}(a, b)(c, d) &= (ac - bd, bc + ad)\\ &= (ca - db, cb + da)\\ &= (c, d)(a, b).\end{aligned}$$

(ii)
$$\begin{aligned}((a, b)(c, d))((e, f)) &= (ac - bd, bc + ad)(e, f)\\ &= ((ac - bd)e - (bc + ad)f,\\ &\qquad (bc + ad)e + (ac - bd)f)\\ &= (a(ce - df) - b(cf + de),\\ &\qquad b(ce - df) + a(cf + de))\\ &= (a, b)((c, d)(e, f)).\end{aligned}$$

(iii) If there exists in $\mathbf{R}^2$ a vector (p, q) such that, for all (a, b) in $\mathbf{R}^2$,

$$(p, q)(a, b) = (a, b),$$

then taking $(a, b) = (1, 0)$, we have

$$(p, q)(1, 0) = (1, 0),$$

i.e.

$$(p, q) = (1, 0),$$

so that $p = 1, q = 0$.

Moreover, a straightforward application of our rule for multiplication yields

$$(1, 0)(a, b) = (a, b) = (a, b)(1, 0).$$

Since $(1, 0) \neq (0, 0)$, we thus have the unique non-zero vector $\mathbf{1} = (1, 0)$, with the required property.

(iv) If $(a, b) \neq 0$ in $\mathbf{R}^2$, then $a^2+b^2 \neq 0$ in $\mathbf{R}$, so (a^2+b^2) has a multiplicative inverse $(a^2+b^2)^{-1}$, written $\dfrac{1}{(a^2+b^2)}$, in $\mathbf{R}$. We then have

$$\left(\frac{a}{(a^2+b^2)}, \frac{-b}{(a^2+b^2)}\right)(a, b) = \left(\frac{(a^2+b^2)}{(a^2+b^2)}, \frac{(ab-ba)}{(a^2+b^2)}\right)$$
$$= (1, 0)$$
$$= \mathbf{1},$$

and

$$(a, b)\left(\frac{a}{(a^2+b^2)}, \frac{-b}{(a^2+b^2)}\right) = \left(\frac{(a^2+b^2)}{(a^2+b^2)}, \frac{(ab-ba)}{(a^2+b^2)}\right)$$
$$= (1, 0)$$
$$= \mathbf{1}.$$

Thus each non-zero vector (a, b) has a multiplicative inverse $\left(\dfrac{a}{(a^2+b^2)}, \dfrac{-b}{(a^2+b^2)}\right)$.

Moreover, if $(a, b) \neq (0, 0)$, and

$$(a, b)(x, y) = (1, 0),$$

we have $$ax - by = 1,$$

and $$bx + ay = 0.$$

Thus $$(a^2+b^2)x = a,$$

and $$(a^2+b^2)y = -b.$$

Since $(a, b) \neq (0, 0)$, it follows that $a^2+b^2 \neq 0$, whence these equations imply that

$$(x, y) = \left(\frac{a}{(a^2+b^2)}, \frac{-b}{(a^2+b^2)}\right),$$

and the given inverses are unique as required.

(v) $$\begin{aligned}(a, b)((c, d)+(e, f)) &= (a, b)(c+e, d+f)\\ &= (a(c+e)-b(d+f),\\ &\qquad b(c+e)+a(d+f))\\ &= ((ac-bd)+(ae-bf),\\ &\qquad (bc+ad)+(be+af))\\ &= ((ac-bd), (bc+ad))\\ &\qquad +((ae-bf), (be+af))\\ &= (a, b)(c, d)+(a, b)(e, f),\end{aligned}$$

and similarly

$$((a, b)+(c, d))(e, f) = (a, b)(e, f)+(c, d)(e, f).$$

(vi) $$\begin{aligned}(\alpha(a, b))(\beta(c, d)) &= (\alpha a, \alpha b)(\beta c, \beta d)\\ &= (\alpha a\beta c - \alpha b\beta d, \alpha b\beta c + \alpha a\beta d)\\ &= (\alpha\beta(ac-bd), \alpha\beta(bc+ad))\\ &= (\alpha\beta)(ac-bd, bc+ad)\\ &= (\alpha\beta)((a, b)(c, d)).\end{aligned}$$

This completes the proof of the theorem.

We should notice that in defining a division algebra over the real field **R**, we have given a particular example of a more general algebraic structure. We do not need to restrict ourselves to the field $\mathbf{R}$. Just as we can have vector spaces over any field, so also we can have division algebras over any field: a vector space over a field F is said to be a *division algebra* over F if there is a rule for multiplication which, in relation to the (scalar) field F, satisfies the rules (i)–(vi) above, just as $\mathbf{R}^2$ does in relation to the real field $\mathbf{R}$.

Worked examples 3.3

1 Compute, in the division algebra $\mathbf{R}^2$, the product $(1, -1)(0, 1)$. Illustrate your answer.

$$(1, -1)(0, 1) = (1, 1)$$

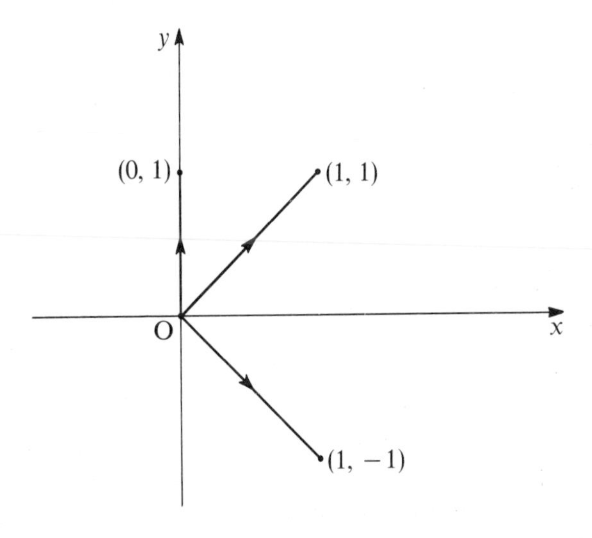

We turn our vector $(1, -1)$ through a right angle by multiplying it by $(0, 1)$.

2 Is it necessary to insist, in the definition of a division algebra, that the vector **1** is unique?

No, since if there were two multiplicative identities

$\mathbf{1}$ and $\mathbf{e}$, they would be such that

$$\mathbf{1u} = \mathbf{u} = \mathbf{u1}$$

$$\mathbf{eu} = \mathbf{u} = \mathbf{ue},$$

for all vectors $\mathbf{u}$. Thus, setting $\mathbf{u} = \mathbf{e}$ and $\mathbf{u} = \mathbf{1}$, respectively,

$$\mathbf{1e} = \mathbf{e} = \mathbf{e1},$$

and

$$\mathbf{e1} = \mathbf{1} = \mathbf{1e}.$$

Thus

$$\mathbf{e} = \mathbf{e1} = \mathbf{1}.$$

Exercises 3.3

1 Compute, in the division algebra $\mathbf{R}^2$, the products

(i) $\left(\frac{1}{\sqrt{2}}, \frac{1}{\sqrt{2}}\right)(0, 1)$,

(ii) $(0, -1)(0, -1)$,

(iii) $(1, 1)(1, -1)$.

Illustrate your answers.

2 Is it necessary to insist, in the definition of a division algebra, that a multiplicative inverse $\mathbf{u}^{-1}$ is unique for any given non-zero vector $\mathbf{u}$?

3 Real numbers of the form $a+b\sqrt{2}$, where a and b are rational numbers, form a vector space over the field of rational numbers (cf. Miscellaneous exercises 2, question 4, page 96). Prove that under real multiplication they form a division algebra over the rational field, i.e. that the multiplication

$$(a+b\sqrt{2})(c+d\sqrt{2}) = (ac+2bd)+(bc+ad)\sqrt{2}$$

satisfies properties (i)–(vi) of the definition of a division algebra, in relation to the rational field.

4 What rules of the real field $\mathbf{R}$ were required in the proofs of properties (i)–(vi) in our last theorem?

3.4 Uniqueness of multiplicative structure in $\mathbf{R}^2$

In the last section we proved that the rule

$$(a, b)(c, d) = (ac - bd, bc + ad)$$

introduces into $\mathbf{R}^2$ a multiplication of vectors which gives $\mathbf{R}^2$ the structure of a division algebra over the real numbers. Since the rules for such a division algebra are identical with the properties of a 'reasonable' multiplication as given in section 3.1, it follows that for this multiplication we know from the work in section 3.2 that there will exist, in $\mathbf{R}^2$, vectors $\mathbf{1}$ and $\mathbf{i}$ such that $\mathbf{i}^2 + \mathbf{1} = 0$. Of course we have already identified the vector $\mathbf{1}$ as the first vector $(1, 0)$ in the (ordered) standard basis for $\mathbf{R}^2$. The vector $\mathbf{i}$ is equally easily identifiable for this multiplication, without having to check through the computations used in section 3.2. For suppose

$$(a, b)^2 = -\mathbf{1} = -(1, 0),$$

then $$(a^2 - b^2, 2ab) = (-1, 0),$$

so that $$a^2 - b^2 = -1,$$

and $$2ab = 0.$$

It follows that $(a, b) = (0, 1)$ or $-(0, 1)$. We therefore choose to denote by $\mathbf{i}$ the second vector $(0, 1)$ in the (ordered) standard basis for $\mathbf{R}^2$.

Thus for this 'standard' multiplication we have

$$\begin{aligned}(a, b) &= a(1, 0) + b(0, 1)\\ &= a\mathbf{1} + b\mathbf{i},\end{aligned}$$

and of course as before the given rule for multiplication

$$(a, b)(c, d) = (ac - bd, bc + ad)$$

can be written in the form

$$(a\mathbf{1} + b\mathbf{i})(c\mathbf{1} + d\mathbf{i}) = (ac - bd)\mathbf{1} + (bc + ad)\mathbf{i}.$$

Given any other 'reasonable' multiplication in $\mathbf{R}^2$, we have of course no reason to suppose that the vectors $\mathbf{1}$ and $\mathbf{i}$ computed as in section 3.2 will be the standard basis vectors (1, 0) and (0, 1) respectively, as they are in this case. What we do know, however, from the work in section 3.2, is that if we are given a multiplication in $\mathbf{R}^2$, which introduces into $\mathbf{R}^2$ the structure of a division algebra over the real numbers, then we can choose a suitable basis for $\mathbf{R}^2$ such that when the vectors in $\mathbf{R}^2$ are referred to this basis, the rule for multiplication is identical with the 'standard' rule. In other words, *by suitable choice of basis, any two real division algebra structures on* $\mathbf{R}^2$ *can be rendered algebraically indistinguishable from one another.*

When $\mathbf{R}^2$ is given the standard real division algebra structure by setting

$$(a, b)(c, d) = (ac-bd, bc+ad),$$

we refer to its vectors as *complex numbers* and use the symbol $\mathbf{C}$ to denote the set of all such numbers. Thus any complex number can be written in the form $x\mathbf{1}+y\mathbf{i}$, which we now write as $x+yi$ (or $x+iy$), omitting the symbol $\mathbf{1}$, the heavy type for $\mathbf{i}$, and regarding the order of i and y as irrelevant. We also usually omit 0 or $0i$ when they occur in a complex number, so that we write x for $x+0i$ and yi for $0+yi$; the complex number $0+0i$ is written 0.

The rule for multiplication is easily remembered, since the formula

$$(x+yi)(x'+y'i) = (xx'-yy')+(yx'-xy')i$$

is obtained by formally multiplying the algebraic expressions $(x+yi)$ *and* $(x'+y'i)$ *in traditional fashion, then replacing* i^2 *by* -1, *wherever it occurs, and collecting together terms not involving* i *and terms involving* i *respectively.*

This usual way of writing complex numbers and calculating products is also useful in division, where we write

$$(x+yi)^{-1} = \frac{1}{x+yi}$$

and proceed as in ordinary algebra subject to the rule $i^2 = -1$, noting in particular that

$$\frac{1}{x+yi} = \frac{x-yi}{(x+yi)(x-yi)} = \frac{x-yi}{x^2+y^2}.$$

Worked examples 3.4

1 Prove that complex numbers of the form $x+0i$ are algebraically indistinguishable from the real numbers.

The rules for addition and multiplication of complex numbers imply that

$$(x+0i)+(y+0i) = (x+y)+0i,$$

and
$$(x+0i)(y+0i) = xy+0i.$$

Moreover the algebraic properties of complex numbers coincide with those of real numbers, i.e. commutativity, distributivity, the existence of an additive identity $0+0i$, of a multiplicative identity $1+0i$, of additive inverses:

$$-(x+0i) = (-x)+0i,$$

and of multiplicative inverses

$$(x+0i)^{-1} = (x^{-1})+0i,$$

are all identical with the respective properties in $\mathbf{R}$, except for the notation $x+0i$ in place of x. Thus algebraically we cannot distinguish between the behaviour of the real numbers and the complex numbers x of the form $x+0i$.

2 Evaluate $(1+i)^5+(1-i)^5$.

$$(1+i)^5 = 1+5i+10i^2+10i^3+5i^4+i^5$$

$$= 1+5i-10-10i+5+i$$

$$= -4-4i.$$

$$(1-i)^5 = 1-5i+10i^2-10i^3+5i^4-i^5$$

$$= 1-5i-10+10i+5-i$$

$$= -4+4i.$$

Thus

$$(1+i)^5+(1-i)^5 = (-4-4i)+(-4+4i)$$

$$= -8+0i$$

$$= -8.$$

3 If the complex number $z = x+yi$ is such that $x^2+y^2 = 1$ and is represented geometrically by a movement $\overrightarrow{OP}$ such that P lies on the circle $x^2+y^2 = 1$, what movement $\overrightarrow{OQ}$ represents iz?

If P moves clockwise around the circle $x^2+y^2 = 1$, what does Q do?

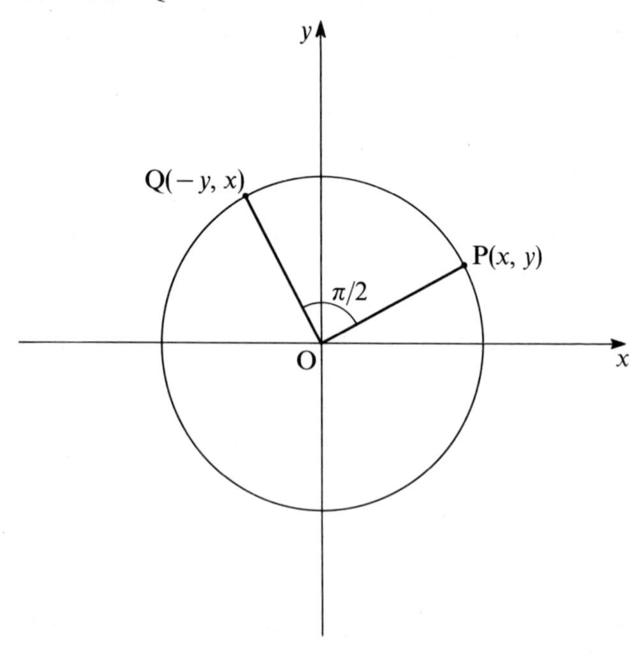

Since $(x+yi)i = (-y)+xi$, and $(-y)^2+x^2 = y^2+x^2 = 1$, Q lies on the circle $x^2+y^2 = 1$. When P is at the point (x, y) Q is at $(-y, x)$, so that $\overrightarrow{OQ}$ is perpendicular to $\overrightarrow{OP}$ and as P moves round the circle Q moves round $\pi/2$ radians behind P, as illustrated on the previous page.

Exercises 3.4

1 Evaluate

(i) $(3-5i)(1+i)$,

(ii) $(\frac{1}{2}+i)(\frac{1}{2}-i)$,

(iii) $\dfrac{1+i}{1-i}$,

(iv) $\dfrac{(3+5i)i}{3-5i}$,

(v) $(1-i)^7$.

2 Evaluate, and illustrate geometrically,

(i) $(1+i)\left(\dfrac{1}{\sqrt{2}}+\dfrac{1}{\sqrt{2}}\right)$,

(ii) $(3+2i)\left(\dfrac{\sqrt{3}}{2}+\dfrac{1}{2}i\right)$,

(iii) $(1+i)^2$,

(iv) $(1+i)^3$.

3 If $\overrightarrow{OP}$ represents a complex number $z = x+yi$, where P moves around the circle $x^2+y^2 = 1$ in a clockwise direction, illustrate and describe the corresponding motion of Q if $\overrightarrow{OQ}$ represents (i) $z+1$, (ii) $z+i$, (iii) $iz+1$.

Miscellaneous exercises 3

1 Express the following in the form $x+yi$:

(i) $\frac{1}{2}(5-4i)+\left(\frac{3-2i}{1+i}\right)$,

(ii) $\dfrac{(1-i)^4}{1+i}$,

(iii) $(a+ib)(a-ib)i \quad (a, b \text{ real})$,

(iv) $(3i)(4-2i)((1+i)-i(1-i))$,

(v) $\dfrac{1+a+ib}{1-a-ib} \quad (a, b \text{ real})$.

2 Find the lengths of the vectors in $\mathbf{R}^2$ corresponding to the following complex numbers. Draw figures illustrating the vectors by means of movements starting at the origin in the real plane.

(i) $1+i\sqrt{3}$,
(ii) $-1-i\sqrt{3}$,
(iii) $1+i$,
(iv) $-3+i$,
(v) $-1-i$.

3 Complex numbers z_1, z_2, z are such that $z = tz_1+(1-t)z_2$, where t is a real number. If $\overrightarrow{\mathrm{OP}}_1$, $\overrightarrow{\mathrm{OP}}_2$ and $\overrightarrow{\mathrm{OP}}$ are displacements in the plane corresponding to the vectors in $\mathbf{R}^2$ associated with z_1, z_2 and z, respectively, what is the geometrical relationship between the points P_1, P_2 and P?

4 A complex number $z = x+yi$ is such that $x^2+y^2 = 1$. What can you say about the complex numbers $-z$, $1+z$, $-iz$? Draw figures in the real plane to illustrate your answers.

5 A multiplication is defined in $\mathbf{R}^2$ by setting

$$(a, b)(c, d) = (ac, bc+ad).$$

Does this multiplication introduce into $\mathbf{R}^2$ the structure of a division algebra over the real numbers?

CHAPTER FOUR

Further algebraic properties of complex numbers

4.1 Absolute values, conjugation, real and imaginary parts

So far we have noted that algebraically the complex numbers $\mathbf{C}$ form a division algebra over the real field. In this chapter we add to this structure two further algebraic properties of $\mathbf{C}$.

The first of these is concerned with what is called the *absolute value*, $|z|$, of a complex number z.

Definition. The *absolute value*, written $|z|$, of the complex number $z = x+yi$ is the (non-negative) real number $\sqrt{(x^2+y^2)}$.

In other words the absolute value $|z|$ of the complex number $z = x+yi$ is precisely the norm $\|(x, y)\|$ of the vector (x, y) in $\mathbf{R}^2$, which corresponds to the complex number z:

$$|x+yi| = \|(x, y)\|.$$

This definition enables us to associate with numbers in $\mathbf{C}$ the geometrical ideas we have already associated with vectors in $\mathbf{R}^2$, concerned with norms interpreted as lengths of displacement vectors. Thus we can say $|z| = \|(x, y)\|$ is the length of the line from the origin O in the plane to the point P(x, y), or that it is the length of the movement from O to P.

In this context we now often call the point $P(x, y)$ the point z, or the point $x+yi$, thus carrying the complex number notation into the real plane. We then speak directly of the distance between two complex numbers z_1, z_2, or the length of the movement from z_1 to z_2. This is just $|z_1-z_2|$, since if $z_1 = x_1+y_1i$ and $z_2 = x_2+y_2i$, we have

$$\begin{aligned}|z_1-z_2| &= |(x_1+y_1i)-(x_2+y_2i)| \\ &= |(x_1-x_2)+(y_1-y_2)i| \\ &= \sqrt{((x_1-x_2)^2+(y_1-y_2)^2)},\end{aligned}$$

which is precisely our Euclidean geometry formula for the distance from (x_1, y_1) to (x_2, y_2), as required.

Since $|x+yi| = \|(x, y)\|$, we can translate the two standard inequalities for norms into complex number form:

$$\begin{aligned}|z_1+z_2| &\leqslant |z_1|+|z_2|, \\ |z_1-z_2| &\geqslant \big||z_1|-|z_2|\big|.\end{aligned}$$

These of course follow immediately from the corresponding inequalities for norms:

$$\begin{aligned}\|\mathbf{u}_1+\mathbf{u}_2\| &\leqslant \|\mathbf{u}_1\|+\|\mathbf{u}_2\|, \\ \|\mathbf{u}_1-\mathbf{u}_2\| &\geqslant \big|\|\mathbf{u}_1\|-\|\mathbf{u}_2\|\big|,\end{aligned}$$

when we set $\mathbf{u}_1$ and $\mathbf{u}_2$ equal to the vectors corresponding to z_1 and z_2 respectively.

The second piece of algebraic structure which we note is that concerned with the operation of *conjugation*. Here we associate with any complex number $z = x+yi$, the complex number $x-yi$.

Definition. The *complex conjugate* of a complex number $z = x+yi$ is denoted by $\bar{z}$ and is the complex number $x-yi$.

We should note immediately that if we repeat the process of conjugation, obtaining $\bar{\bar{z}}$ from $\bar{z}$, we recover the complex number from which we started, i.e. $\bar{\bar{z}} = z$ for all z in $\mathbf{C}$. For this reason we say conjugation is an *involution*.

We may also note the connection of conjugation with addition and multiplication in **C**. Thus

$$\overline{z_1+z_2} = \bar{z}_1+\bar{z}_2$$

and
$$\overline{z_1 z_2} = \bar{z}_1\bar{z}_2.$$

The proofs of these facts are easily verified using the definition $\overline{x+yi} = x-yi$.

Conjugation and absolute values can be agreeably mixed. Thus if $z = x+yi$

$$\begin{aligned} z\bar{z} &= (x+yi)(x-yi) \\ &= x^2+y^2 \\ &= |z|^2. \end{aligned}$$

This result has two immediately useful applications. The first concerns the relation between absolute values and multiplication. Thus

$$\begin{aligned} (|z_1|\,|z_2|)^2 &= |z_1|^2\,|z_2|^2 \\ &= z_1\bar{z}_1 z_2\bar{z}_2 \\ &= z_1 z_2\bar{z}_1\bar{z}_2 \\ &= (z_1 z_2)(\overline{z_1 z_2}) \\ &= |z_1 z_2|^2. \end{aligned}$$

Since $|z_1|$, $|z_2|$ and $|z_1 z_2|$ are non-negative real numbers, it follows that

$$|z_1|\,|z_2| = |z_1 z_2|.$$

The second application is in connection with inverses of complex numbers. The cumbersome formula

$$(x+yi)^{-1} = \frac{x}{x^2+y^2} - \frac{y}{x^2+y^2}i$$

can now be written

$$z^{-1} = \frac{\bar{z}}{|z|^2},$$

for if $z \neq 0$, it follows that $|z| \neq 0$, and the formula

$$z\bar{z} = |z|^2$$

implies
$$z\left(\frac{\bar{z}}{|z|^2}\right) = 1,$$

i.e.
$$z^{-1} = \frac{\bar{z}}{|z|^2}.$$

(In particular, if $|z| = 1$, $z^{-1} = \bar{z}$.)

Finally, for convenience, at this point we introduce two further pieces of notation. If $z = x+yi$, we refer to x as the *real part* of z and write $\mathrm{Rl}(z) = x$, and to y as the *imaginary part* of z, writing $\mathrm{Im}(z) = y$. The use of the word *real* in this context refers simply to the fact that complex numbers of the form $x+0i$ behave like real numbers. The use of the word *imaginary* is historical: as we have seen, complex numbers are not 'imaginary' or 'unreal'; they come from perfectly 'real' vectors.

These real and imaginary parts are readily related to conjugates of complex numbers. Thus clearly from our definitions, we have

$$\mathrm{Rl}(z) = \frac{z+\bar{z}}{2},$$

and
$$\mathrm{Im}(z) = \frac{z-\bar{z}}{2}.$$

Worked examples 4.1

1 Describe geometrically the set of points (x, y) in the real plane which corresponds to the complex numbers $z = x+yi$ such that $|z-i| < 2$.

Since $|z-i|$ is the distance from the 'point' z to the 'point' i, to require $|z-i| < 2$ is to ask that we restrict our attention to points whose distance from the point i is less than 2. In other words we should consider only points inside a circle centre $i = (0, 1)$, radius 2. Algebraically w have

$$\begin{aligned} |z-i| &= |x+yi-i| \\ &= |x+(y-1)i| \\ &= \sqrt{((x^2+(y-1)^2)}. \end{aligned}$$

Thus $|z-i| < 2$ if and only if $x^2+(y-1)^2 < 4$, i.e. z is restricted to lie in the interior of the circle $x^2+(y-1)^2 = 4$.

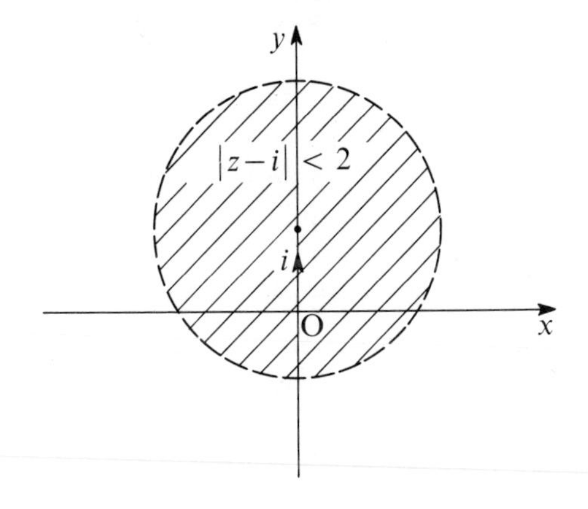

2 If $|z+1| \leqslant 3$, what are the greatest and least values of $|z+2|$?

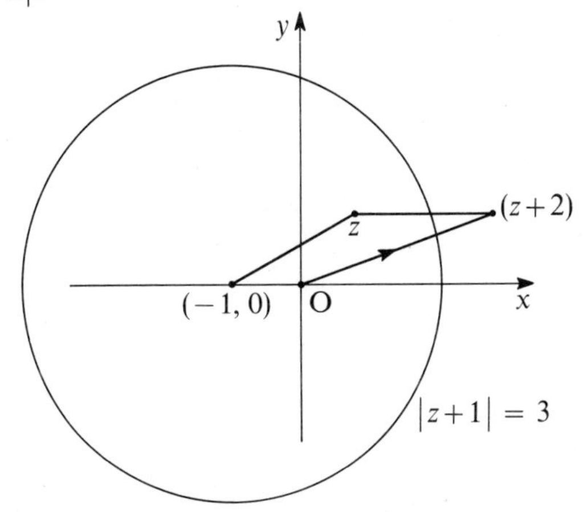

Here we know, since $|z+1| \leqslant 3$, that z must lie at most a distance 3 from the point -1. Thus z is restricted to lying on or in the circle radius 3, centre $(-1, 0)$. We require to find the greatest and least possible distances of $z+2$ from 0, given this condition.

From the diagram it is intuitively clear that the greatest value of $|z+2|$, i.e. the greatest value of the length of the line OP, is when OP lies along the x-axis and z is on the circle $|z+1| = 3$, i.e. when the length of OP is 4. The least value is also clearly 0.

Algebraically we prove this is as follows. On the one hand

$$\begin{aligned} |z+2| &= |z+1+1| \\ &\leqslant |z+1|+|1| \\ &\leqslant 3+1 \\ &= 4, \end{aligned}$$

and the value 4 is attained when $z = 2$. On the other hand, since $|z+2|$ is never negative, $|z+2| \geqslant 0$. Taking $z = -2$, we have $|z+2| = 0$ and $|z+1| \leqslant 3$, so that again this lower bound, 0, is attained.

3 Prove that if c is a real number, then for any complex number b such that $|b|^2 > c$,

$$z\bar{z}+\bar{b}z+b\bar{z}+c = 0$$

is the equation of a circle. Find the centre of the circle and its radius in terms of b and c.

We have

$$\begin{aligned} z\bar{z}+\bar{b}z+b\bar{z}+c &= (z+b)(\bar{z}+\bar{b})-b\bar{b}+c \\ &= (z+b)(\overline{z+b})-b\bar{b}+c \\ &= |z+b|^2-|b|^2+c. \end{aligned}$$

Our equation can thus be written

$$|z+b|^2 = |b|^2-c,$$

and therefore represents a circle, centre $-b$ and radius $\sqrt{(|b|^2-c)}$.

Exercises 4.1

1 Find the real part, the imaginary part, the absolute value and the conjugate of the following complex numbers.

(i) $(i-1)^2/(i+1)$,

(ii) $i\overline{(a+ib)}$,

(iii) $|(1+i)|(1-i)$.

2 Prove the equations given on page 123:

$$\overline{z_1+z_2} = \bar{z}_1-\bar{z}_2,$$

$$\overline{z_1z_2} = \bar{z}_1\bar{z}_2.$$

3 Describe geometrically the sets of points (x, y) in the real plane corresponding to complex numbers $z = x+yi$,

(i) such that $|z| \leqslant 1$,

(ii) such that $|z-3| = 2$,

(iii) such that $\mathrm{Rl}\, z \leqslant 0$,

(iv) such that $|z| \leqslant 1$ and $\mathrm{Im}\,(z) \geqslant 0$.

4 Prove that if $|z| \leqslant 1$, then $1 \leqslant |iz-2| \leqslant 3$.

5 Prove that for all complex numbers z_1, z_2

$$|z_1+z_2|^2+|z_1-z_2|^2 = 2(|z_1|^2+|z_2|^2).$$

Interpret the equation geometrically.

6 Prove that, if q is a real number, then for any non-zero complex number p,

$$\bar{p}z+p\bar{z} = q$$

is the equation of a straight line. Find the gradient of the line in terms of p.

7 Which of the following equations are true?
(i) $\mathrm{Rl}(z_1+z_2) = \mathrm{Rl}(z_1)+\mathrm{Rl}(z_2)$,
(ii) $\mathrm{Rl}(z_1 z_2) = \mathrm{Rl}(z_1)\,\mathrm{Rl}(z_2)$,
(iii) $\mathrm{Rl}(z_1 z_2) = \mathrm{Rl}(z_1)\,\mathrm{Rl}(z_2)-\mathrm{Im}(z_1)\,\mathrm{Im}(z_2)$,
(iv) $\mathrm{Rl}(az) = a\,\mathrm{Rl}(z)$ if a is real.

4.2 The algebraic closure of the field of complex numbers

Apart from noting the newer properties of absolute value and conjugation given in the last section, we may well ask if the complex numbers **C** will further extend our earlier algebraic procedures used in dealing with fields of numbers. We therefore now compare the algebraic behaviour of **C** with that of the field of real numbers, and the field of rational numbers.

First we ask whether **C** is a field. We have proved that it is a division algebra over the real field; what properties do division algebras and fields have in common? In both, we have a commutative, associative process of addition, in which there is a zero and in which we have additive inverses. Again, in both we have a commutative, associative process of multiplication, a multiplication identity, and multiplicative inverses for all non-zero objects. The two processes of addition and multiplication are related by distributive rules in both cases. In fact, if the reader cares to compare the precise definitions of the two algebraic structures (field on p. 45, division algebra on p. 109), it is clear that they are identical, except for the fact that in the case of a division algebra we have scalar multiplication. Thus if we 'forget' scalar multiplication the complex numbers **C** form a field.†

In forming **C**, we have of course in a very reasonable sense 'enlarged' **R**, for example 1 of Worked examples 3.4 of the last chapter showed that complex numbers of the form

†The observant reader may well now claim that in considering **C** as a field we have not really 'lost' the scalar multiplication given to us when we consider **C** as a division algebra, for multiplication in **C** by 'real numbers' $\alpha+0i$, yields $(\alpha+0i)(x+yi) = \alpha x+(\alpha y)i$, i.e. our original scalar multiplication $\alpha(x, y) = (\alpha x, \alpha y)$ in $\mathbf{R}^2$.

$x+0i$ could not be distinguished algebraically from the real numbers. This 'enlargement' enables us to solve the equation

$$z^2+1 = 0,$$

which we saw in Exercise 2 of Miscellaneous exercises 1 was not soluble in the real field. For we now write this equation

$$(x+yi)^2+(1+0i)^2 = 0+0i,$$

and deduce that

$$x^2-y^2+1+2xyi = 0+0i,$$

i.e.
$$x^2-y^2+1 = 0,$$

and
$$2xy = 0.$$

Thus either $x = 0$, or $y = 0$ (but not both, since this would imply, from the first equation, that $1 = 0$). If $y = 0$, then x is a real number such that $x^2+1 = 0$, which we have already observed is impossible. Thus $x = 0$ and $y^2 = 1$, i.e. $y = \pm 1$, and $z^2+1 = 0$ has two solutions $0+1i$ and $0+(-1)i$, i.e. i and $-i$, in $\mathbf{C}$.

We can thus view our construction of $\mathbf{C}$ as another step in the enlargement of our fields of numbers, at each stage of which we were able to solve equations not soluble previously: over the rational field $x^2 = 2$ is not soluble, but it is over the real field; over the real field $x^2+1 = 0$ is not soluble, but it is over the complex field. It is reasonable to ask whether this process continues, i.e. do there exist equations

$$a_n z^n + a_{n-1}z^{n-1} + \ldots + a_1 z + a_0 = 0$$

with complex coefficients a_0, a_1, ..., a_n, and no complex number solutions? A proof of the answer 'no' to this question is beyond the scope of this book, but we state it for completeness: *the field* $\mathbf{C}$ *of complex numbers is an algebraically closed field, i.e. every polynomial equation of* nth *degree with complex coefficients has precisely* n *complex solutions (some of which may be 'repeated')*. In particular therefore an nth degree polynomial equation with real coefficients, when regarded as an equation over the complex field, will have n solutions, possibly repeated, *all in the field* $\mathbf{C}$ *of complex numbers.* We may observe that in this case we have the following result.

Proposition. *If the complex number* $a+bi$ *is a solution of the complex polynomial equation* $f(z) = a_0+a_1z+a_2z^2+\ldots+a_nz^n = 0$ *in which* $a_0 = a_0+0i$, $a_1 = a_1+0i$, ..., $a_n = a_n+0i$ *are real numbers in* $\mathbf{C}$, *then the conjugate* $a-bi$ *is also a solution of the equation.*

Proof. Let $\alpha = a+bi$. Then

$$f(\alpha) = a_0+a_1\alpha+\ldots+a_n\alpha^n = 0,$$

so that

$$\begin{aligned} f(\bar{\alpha}) &= a_0+a_1\bar{\alpha}+\ldots+a_n\bar{\alpha}^n \\ &= \bar{a}_0+\bar{a}_1\bar{\alpha}+\ldots+\bar{a}_n\bar{\alpha}^n \\ &= \bar{a}_0+\overline{a_1\alpha}+\ldots+\overline{a_n\alpha^n} \\ &= \overline{a_0+a_1\alpha+\ldots+a_n\alpha^n} \end{aligned}$$

(since the numbers a_i are real)

$$\begin{aligned} &= \overline{f(\alpha)} \\ &= \bar{0} \\ &= 0, \end{aligned}$$

and $\bar{\alpha}$ is a root of the equation $f(z) = 0$, as required.

Worked examples 4.2

1 If α, β are complex numbers such that $\alpha \neq 0$, solve the equation $\alpha z+\beta = 0$ in $\mathbf{C}$.

If $\alpha = a+bi$, $\beta = c+di$, find the necessary and sufficient condition for the equation to have a real root, in terms of a, b, c and d.

Since $\alpha \neq 0$, $\alpha^{-1} = 1/\alpha$ exists in **C** and clearly $z = -\beta/\alpha$ is the required solution.

There will thus be a real root if and only if $\alpha^{-1}\beta$ is real. But

$$\alpha^{-1}\beta = \frac{c+di}{a+bi}$$

$$= \frac{(c+di)(a-bi)}{a^2+b^2}$$

$$= \frac{(ca+db)}{a^2+b^2} + \frac{(ad-bc)i}{a^2+b^2},$$

thus the root is real if and only if $\dfrac{ad-bc}{a^2+b^2} = 0$, i.e. if and only if $ad = bc$.

2 Solve the complex equation

$$z^2+\alpha z+\beta = 0,$$

and write out the solutions in cases

(*a*) $\alpha = 0, \beta = -i$,
(*b*) $\alpha = \beta = 1$.

'Completing the square' throws the given equation into the form

$$\left(z+\frac{\alpha}{2}\right)^2 = \frac{\alpha^2}{4}-\beta.$$

We therefore solve the equation by solving the complex equation

$$w^2 = \gamma$$

and then setting $z+\alpha/2 = w$, and $\alpha^2/4-\beta = \gamma$.

Suppose therefore that $\gamma = a+bi$. We require to find real numbers x and y such that

(i) $$(x+yi)^2 = a+bi,$$

i.e. such that

(ii) $$x^2+y^2 = a$$

and

(iii) $$2xy = b.$$

Thus, if such real numbers x and y exist, we require

$$\begin{aligned}(x^2+y^2)^2 &= x^4+2x^2y^2+y^4\\ &= x^4-2x^2y^2+y^4+4x^2y^2\\ &= (x^2-y^2)^2+(2xy)^2,\\ &= a^2+b^2,\end{aligned}$$

so that

(iv) $$x^2+y^2 = \sqrt{(a^2+b^2)}.$$

Combining equations (ii) and (iv), we have

$$x^2 = \tfrac{1}{2}(a+\sqrt{(a^2+b^2)}),\ y^2 = \tfrac{1}{2}(-a+\sqrt{(a^2+b^2)}).$$

Clearly we can find x and y in the real field **R** satisfying these equations, and in so doing we must choose signs in such a way that equation (iii) is satisfied. Thus in case $b > 0$ we take x and y with the same sign; in case $b < 0$ we take x and y with opposite signs. There are thus at most two solutions of the equation $w^2 = \gamma$, and an easy check reveals that both are indeed solutions, so that there are exactly two solutions of $w^2 = \gamma$ and hence of the given equation.

Now we deal with the special cases.

(*a*) Here we have $a = 0$, $b = 1$ so $x^2 = y^2 = \frac{1}{2}$ and the solutions of $z^2 = i$ are

$$z = \pm\frac{1}{\sqrt{2}}(1+i)$$

(*b*) In this case we simply have the real equation $z^2+z+1 = 0$, with solutions

$$z = \tfrac{1}{2}(1\pm i\sqrt{3}).$$

Exercises 4.2

1 Solve the complex equations
(i) $z^2 = 5+6i$,
(ii) $z^2-2z+6 = 0$,
(iii) $z^2+z+(1-i) = 0$.

2 Prove that the complex equation

$$z^2+\alpha z+\beta = 0$$

has equal roots if $\alpha^2 = 4\beta$.

3 Prove that if the complex numbers $\gamma = a+bi$ and its conjugate $\bar{\gamma} = a-bi$ are the roots of the equation

$$z^2+\alpha z+\beta = 0,$$

then α and β are real numbers.

4.3 Ordering in the complex field

Having recognized one algebraic difference between the complex number field $\mathbf{C}$ and the fields of real and rational numbers, we now consider another vital difference. We prove that the complex field cannot be made into an ordered field, despite the fact that it contains the real field and the rational field, both of which are ordered fields.

Let us repeat the conditions which must be satisfied if $\mathbf{C}$ is to be an ordered field: the complex number field $\mathbf{C}$ will constitute an ordered field if it contains a non-empty collection of complex numbers, $\mathbf{C}^+$, which we would call *positive*, such that

(i) for all z in $\mathbf{C}^+$ one and only one of the following relations holds: z is in $\mathbf{C}^+$, $z = 0$, or $(-z)$ is in $\mathbf{C}^+$;

(ii) for all w, z in $\mathbf{C}^+$, $w+z$ and wz are in $\mathbf{C}^+$.

Proposition. *The complex field* $\mathbf{C}$ *cannot be ordered.*

Proof. We prove that no choice of $\mathbf{C}^+$ can be made, satisfying the conditions (i) and (ii) above, without contradiction.

Consider therefore the complex number i in $\mathbf{C}$. If $\mathbf{C}$ is ordered, by condition (i) of our rules for an ordered field, i is either in $\mathbf{C}^+$, or $i = 0$, or $-i$ is in $\mathbf{C}^+$. By definition, $i \neq 0$, so suppose i is in $\mathbf{C}^+$. Then by the second of our conditions for ordering, $ii = i^2 = -1$ is in $\mathbf{C}^+$. But then, again by condition (ii), $(-1)i = -i$ is also in $\mathbf{C}^+$, which contradicts condition (i), since i and $-i$ cannot both be in $\mathbf{C}^+$. So assuming i is in $\mathbf{C}^+$ implies it is not in $\mathbf{C}^+$; contradiction!

Assume therefore that $-i$ is in $\mathbf{C}^+$. Then $(-i)(-i) = i^2 = -1$ is in $\mathbf{C}^+$, whence $(-1)(-i) = i$ is in $\mathbf{C}^+$, which again contradicts condition (ii), since if $-i$ is in $\mathbf{C}^+$ as we assumed, then $-(-i) = i$ is not in $\mathbf{C}^+$.

Thus we cannot define *positive* complex numbers $\mathbf{C}^+$, and no ordering of the field of complex numbers is possible: as a consequence we cannot speak of one complex number being *greater than* another. It is this lack of ordering which renders the use of absolute values, real parts and imaginary parts all the more useful. For these enable us to pass to the real field where we can use our theory of real inequalities.

Miscellaneous exercises 4

1 Solve the complex equations
(i) $z^2 - 2(1+i)z + (3-2i) = 0$,
(ii) $z^3 + z^2 + z + 1 = 0$.

2 If $|z| = 1$, describe geometrically the points in the real plane corresponding to the complex number $z + 1/z$.

3 Describe geometrically the set of points (x, y) in the real plane, corresponding to the complex numbers $z = x + yi$, such that

$$|z-1| + |z+1| \leqslant 3.$$

4 If $|z-1| = 1$, prove that the points in the real plane corresponding to the complex number $w = 1/z$ lie on a straight line perpendicular to the x-axis passing through the point $(\frac{1}{2}, 0)$.

CHAPTER FIVE

Geometrical interpretations of multiplication in $\mathbf{C}$

5.1 Multiplication in $\mathbf{C}$ and rotations of $\mathbf{R}^2$

Since addition of complex numbers is defined by means of vector addition in the real vector space $\mathbf{R}^2$ which underlies $\mathbf{C}$, it has been clear from the start of our study of $\mathbf{R}^2$ and $\mathbf{C}$ how to interpret geometrically the addition of complex numbers. However when we turn to multiplication of complex numbers:

$$(a+bi)(c+di) = (ac-bd)+(bc+ad)i,$$

it is not at all as clear what geometrical interpretation, if any, we can make.

In fact, before we can start making such a geometrical interpretation we must add further algebraic structure to $\mathbf{R}^2$. This extra structure which we require is concerned with what we shall call *rotation mappings* of the vector space $\mathbf{R}^2$. It would be easy just to make an algebraic definition of such a 'mapping' and to proceed immediately from there, but to do so would not be particularly enlightening from a geometrical point of view, at least not without a considerable act of faith on the part of the reader. We therefore choose first to recall in an informal manner how we deal geometrically with rotations in the co-ordinate geometry of the real plane. We then consider how we can use these ideas to formulate suitable formal and purely algebraic definitions in $\mathbf{R}^2$, using only the vector space rules which define $\mathbf{R}^2$.

Consider therefore the usual derivation of the formulae used in co-ordinate geometry to relate the co-ordinates of a point P with those of a point P′ to which P moves under a rotation of the co-ordinate plane through an angle θ about the origin. The reader may care to recall that we can picture the situation by thinking of the plane as a piece of paper covered by a piece of transparent paper on which the axes and co-ordinates are drawn. We then pin down the two pieces of paper through the origin; hold the transparent paper (and hence the axes) fixed, and rotate the 'plane', about the pinned origin, under the transparent paper. The usual diagram given in this situation is of course as follows.

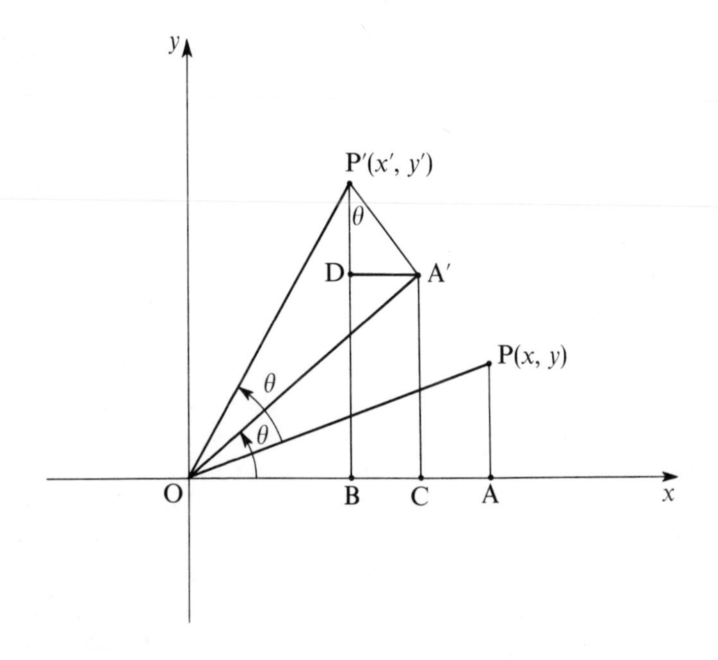

In this diagram OA rotates to OA′ as OP rotates to OP′; $\angle \mathrm{PAO} = \angle \mathrm{P'A'O} = \pi/2$; P′B and A′C are perpendicular to Ox, and A′D is perpendicular to P′B. If P has co-ordinates (x, y) and P′ co-ordinates (x', y'), we have

$$\begin{aligned} x' &= \mathrm{OB} \\ &= \mathrm{OC}-\mathrm{BC} \\ &= \mathrm{OC}-\mathrm{DA'} \\ &= \mathrm{OA'}\cos\theta-\mathrm{P'A'}\sin\theta \\ &= \mathrm{OA}\cos\theta-\mathrm{PA}\sin\theta \\ &= x\cos\theta-y\sin\theta. \end{aligned}$$

Similarly

$$\begin{aligned} y' &= \mathrm{P'B} \\ &= \mathrm{P'D}+\mathrm{DB} \\ &= \mathrm{P'D}+\mathrm{A'C} \\ &= \mathrm{P'A}\cos\theta+\mathrm{OA'}\sin\theta \\ &= \mathrm{PA}\cos\theta+\mathrm{OA}\sin\theta \\ &= y\cos\theta+x\sin\theta. \end{aligned}$$

We thus obtain the formulae

$$\begin{aligned} x' &= x\cos\theta-y\sin\theta \\ y' &= x\sin\theta+y\cos\theta, \end{aligned}$$

which determine in co-ordinate terms a rotation of the real plane through an angle θ, by giving the co-ordinates (x', y') of any point P′ in terms of the co-ordinates (x, y) of the point P which moves to P′ under the rotation.

It should be said that the above derivation is not complete, because it only deals with the case when P and P′ lie in the first quadrant, and in other cases there has sometimes to be a careful adjustment of signs. We leave it as an exercise to the reader to write out a proof for some other case, for example when P lies in the first quadrant but P′ lies in the second quadrant.

However, what we shall do is accept these formulae as

they stand and use them to suggest an algebraic definition of a rotation mapping of the vector space $\mathbf{R}^2$, which will be independent of this geometrical justification. Since the vector space rules for $\mathbf{R}^2$ contain no mention of angles, nor of sines and cosines of angles, we begin by rewriting the formulae without use of these trigonometrical functions, in the hope that by so doing we shall obtain formulae more readily useful in $\mathbf{R}^2$. This is not particularly difficult if we note that a rotation of our paper 'plane' can be specified without mentioning the angle θ, by fixing our attention on one particular point P and stating which point P′ we require it to be moved to by the rotation. In co-ordinate, rather than physical terms, if we take P with co-ordinates (1, 0) and P′ with co-ordinates (a, b):

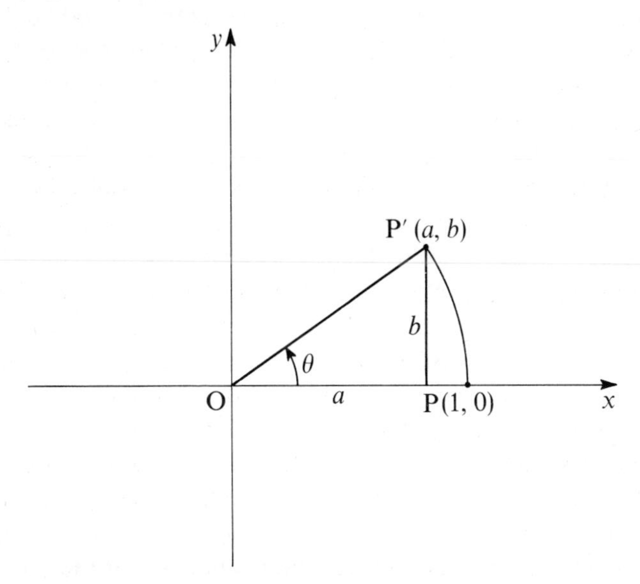

we have $\cos\theta = a$, $\sin\theta = b$, and of course $a^2+b^2 = 1$. Without using trigonometrical functions we can therefore specify the rotation by the formulae

$$x' = ax-by$$

$$y' = bx+ay,$$

where $a^2+b^2 = 1$. Having now got these formulae in this form, without sines and cosines, we can immediately use them in $\mathbf{R}^2$ to define a rotational mapping which carries a *vector* (x, y) into a *vector* (x', y').

Definition. Given an ordered pair of real numbers (a, b) such that $a^2+b^2 = 1$, the *rotation mapping* $\rho_{(a, b)}$ of the vector space $\mathbf{R}^2$, associated with (a, b), is the transformation of vectors which assigns to any vector $\mathbf{u} = (x, y)$ in $\mathbf{R}^2$ the vector $\rho_{(a, b)}(\mathbf{u}) = (ax-by, bx+ay)$ in $\mathbf{R}^2$.

This then is the algebraic definition with which we could have begun our discussion of rotations of $\mathbf{R}^2$, avoiding all reference to the associated geometrical situation. We might now ask the reader to forget, or at least ignore, his knowledge of 'turning about a point' in the plane, of the 'angle' associated with such a turn, or of the theory of sines and cosines in elementary trigonometry, and to proceed from now on to deal with rotations only in terms of rotation mappings in the algebraic context of the vector space $\mathbf{R}^2$, working on the basis of this definition. Such a request would surely be unrealistic; certainly we now have a sound algebraic definition for use in $\mathbf{R}^2$, but we also have an associated geometrical interpretation, and in dealing with this interpretation *in an informal way* we assume the reader will, for the time being at least, make use of any suitable geometrical or trigonometrical methods. This section ends with worked examples illustrating this point.

We shall later show how to extend our algebraic theory of $\mathbf{R}^2$ so that full and formal use of standard trigonometrical theory can be made *in the vector space* $\mathbf{R}^2$, and not just informally in our geometrical interpretation. In the meantime however, let us at least see how our definition of a rotation mapping of $\mathbf{R}^2$ gives the geometrical interpretation of the multiplication of complex numbers for which we are looking.

Using complex multiplications of vectors, the definition can be written in the form

$$\rho_{(a, b)}(x, y) = (a, b)(x, y).$$

With this formula we can immediately interpret complex multiplication, particularly in the case of complex numbers of unit absolute value. For given such a complex number $a+bi$, with $a^2+b^2 = 1$, we have available an ordered pair of real numbers (a, b) such that $a^2+b^2 = 1$. We thus have a rotation $\rho_{(a,b)}$ and our formula now indicates that multiplication of any complex numbers $x+yi$ simply rotates the vector (x, y), associated with $x+yi$ in $\mathbf{R}^2$, by means of the rotation mapping $\rho_{(a,b)}$ determined by the ordered pair (a, b). An extension of this to include multiplication by any non-zero complex number $c+di$, whatever its absolute value, can also be made. For given $c+di$, we associate with it $a+bi$ where $a = c/\sqrt{(c^2+d^2)}$ and $b = d/\sqrt{(c^2+d^2)}$, so that $a^2+b^2 = 1$. Multiplication by $c+di$:

$$(c+di)(x+yi) = \{\sqrt{(c^2+d^2)}\}(a+bi)(x+yi)$$

then involves firstly the rotation of (x, y) by $\rho_{(a,b)}$ and secondly scalar multiplication of $(a, b)(x, y)$ by the factor $\sqrt{(c^2+d^2)}$. *Taking the usual pictorial representation of* $\mathbf{R}^2$ *we can thus interpret complex multiplication by rotations and 'magnifications' of the real plane.* Typical diagrams are as follows.

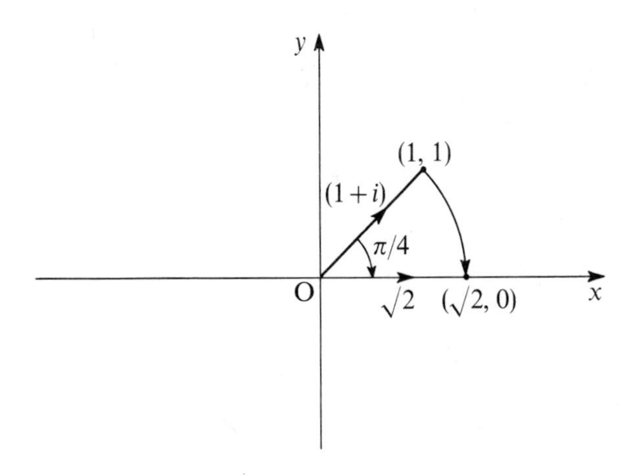

Multiplication of $1+i$ by $\frac{1}{\sqrt{2}}-\frac{i}{\sqrt{2}}$, giving $\sqrt{2}+0i$.

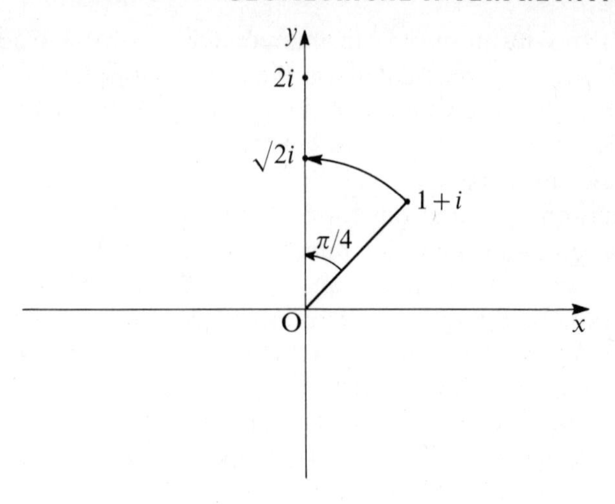

Multiplication of $1+i$ by $1-i$, giving $2i$.

Worked examples 5.1

1 Determine the rotation mapping of $\mathbf{R}^2$ which corresponds to a geometrical rotation of the co-ordinate plane about the origin through an angle $2\pi/3$ (i) clockwise, (ii) anticlockwise.

(i) Under a clockwise rotation through $2\pi/3$ the point $(1, 0)$ moves to the point $\left(-\frac{1}{2}, -\frac{\sqrt{3}}{2}\right)$:

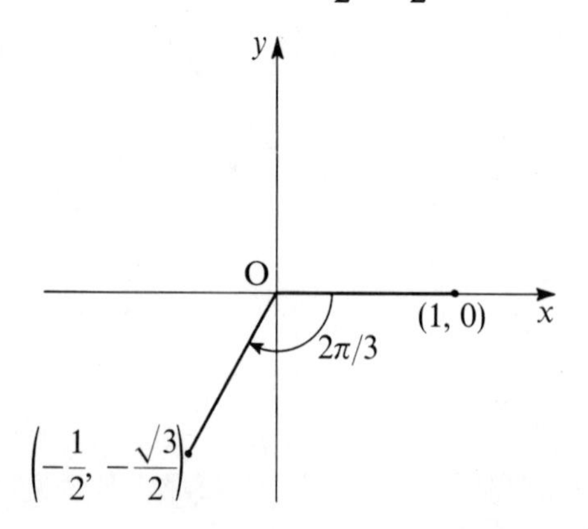

The appropriate rotation mapping of $\mathbf{R}^2$ is ρ, where

$$\rho(x, y) = \left(-\frac{1}{2}x + \frac{\sqrt{3}}{2}y, -\frac{\sqrt{3}}{2}x - \frac{1}{2}y\right).$$

(ii) If the rotation is anticlockwise, the point $(1, 0)$ moves to $-\frac{1}{2}, \frac{\sqrt{3}}{2}$:

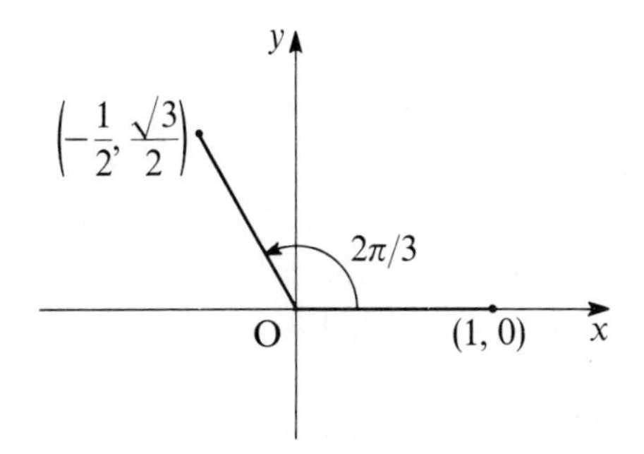

and the rotation mapping is ρ, where

$$\rho(x, y) = \left(-\frac{1}{2}x - \frac{\sqrt{3}}{2} \frac{\sqrt{3}}{2}x - \frac{1}{2}y\right)$$

2 By considering what happens to the vector $(1, 0)$, interpret geometrically the effect of multiplying complex numbers by (i) i, (ii) $-i$, (iii) $1+i$.

(i) Since $i(x+yi) = -y+xi$, multiplication by i corresponds to the rotation mapping ρ given by $\rho(x, y) = (-y, x)$. Thus multiplication by i rotates $(1, 0)$ to $(0, 1)$, i.e. rotates the plane through a right angle in an anticlockwise direction.

(ii) In this case we have $-i(x+yi) = y-xi$, so $(1, 0)$ is rotated to $(0, -1)$, i.e. multiplication by $-i$ rotates the plane through a right angle in a clockwise direction.

(iii) We have $1+i = \sqrt{2}\left(\frac{1}{\sqrt{2}} + \frac{i}{\sqrt{2}}\right)$, and

$$\left(\frac{1}{\sqrt{2}}+\frac{i}{\sqrt{2}}\right)\left(x+yi\right)=\left(\frac{x}{\sqrt{2}}-\frac{y}{\sqrt{2}}\right)+i\left(\frac{x}{\sqrt{2}}+\frac{y}{\sqrt{2}}\right).$$

Thus multiplication by $1+i$ rotates $(1,0)$ to $\left(\frac{1}{\sqrt{2}},\frac{1}{\sqrt{2}}\right)$ and then magnifies the vector to obtain $(1,1)$, i.e. rotates the plane anticlockwise through an angle $\pi/4$, and then 'magnifies' the plane by a factor $\sqrt{2}$.

Exercises 5.1

1 Determine the rotation mapping of $\mathbf{R}^2$ which corresponds to a rotation of the co-ordinate plane about the origin (i) clockwise through π, (ii) anticlockwise through $\pi/3$, (iii) clockwise through $\pi/4$.

2 Interpret geometrically the effect of multiplying complex numbers (i) by $1-i$, (ii) by $2i$, (iii) by $\cos\alpha+i\sin\alpha$.

5.2 The algebraic structure of rotation mappings of $\mathbf{R}^2$

By definition, a rotation mapping $\rho_{(a,b)}$ of $\mathbf{R}^2$ is determined by multiplying vectors in $\mathbf{R}^2$ by the vector (a,b) according to our rules for complex multiplication: for all (x,y) in $\mathbf{R}^2$, we have

$$\rho_{(a,b)}(x,y) = (a,b)(x,y).$$

It follows that if we are given two rotation mappings $\rho_{(c,d)}$ and $\rho_{(a,b)}$, and apply to $\mathbf{R}^2$ first the one and then the other in the given order, we shall multiply vectors in $\mathbf{R}^2$ first by (c,d) and then by (a,b):

$$\begin{aligned}\rho_{(a,b)}(\rho_{(c,d)}(x,y)) &= \rho_{(a,b)}((c,d)(x,y))\\ &= (a,b)((c,d)(x,y)).\end{aligned}$$

But (by associativity)

$$(a,b)((c,d)(x,y)) = ((a,b)(c,d))(x,y),$$

and since

$$(a, b)(c, d) = (ac-bd, bc+ad),$$

it follows that

$$\begin{aligned}\rho_{(a,b)}(\rho_{(c,d)}(x, y) &= (ac-bd, bc+ad)(x, y) \\ &= \rho_{(ac-bd,\, bc+ad)}(x, y).\end{aligned}$$

Thus the application of one rotation mapping followed by another yields a third rotation mapping determined by the product of the complex numbers associated with the given rotation mappings. Since complex multiplication is commutative, the order in which we apply two successive rotation mappings is irrelevant.

When we follow one rotation mapping by another, we speak of *composing* the mappings. Given two rotation mappings $\rho_{(a,b)}$ and $\rho_{(c,d)}$ as above, we therefore speak of their *composition* and write it

$$\rho_{(a,b)} \circ \rho_{(c,d)}$$

intending $\rho_{(c,d)}$ to be first applied and $\rho_{(a,b)}$ second. What we have verified here is that this composition is again a rotation mapping, given by

$$\rho_{(a,b)} \circ \rho_{(c,d)} = \rho_{(ac-bd,\, bc+ad)}.$$

In these terms we may observe that the multiplicative algebraic structure of complex numbers of unit absolute value is reflected in an algebra of composition of rotation mappings of $\mathbf{R}^2$. On the complex number side we have a commutative associative multiplication, any two complex numbers of unit absolute value determining a third number also of unit absolute value; we have an 'identity', namely $1 = 1+0i$ of unit absolute value, which leaves any complex number unchanged when it is used as a multiplier; we also have associated with any complex number $a+bi$ of unit absolute value a multiplicative inverse $a-bi$, also of unit absolute value, for

$$\begin{aligned}(a-bi)(a+bi) &= a^2+b^2 \\ &= 1\end{aligned}$$

On the rotational side we have a commutative associative composition of rotation mappings of $\mathbf{R}^2$. We have an 'identity' mapping, $\rho_{(1,0)}$ corresponding to the complex number $1+0i$, which leaves all vectors in $\mathbf{R}^2$ fixed when it is applied to them:

$$\begin{aligned}\rho_{(1,0)}(x, y) &= (1,0)(x, y)\\ &= (x, y).\end{aligned}$$

Finally, given a rotation mapping $\rho_{(a,b)}$ we associate with it the 'inverse' mapping $\rho_{(a,-b)}$ (which in geometrical terms is the rotation equal to $\rho_{(a,b)}$ but in the opposite sense), such that

$$\begin{aligned}\rho_{(a,b)} \circ \rho_{(a,-b)} &= \rho_{(a^2+b^2,0)}\\ &= \rho_{(1,0)}.\end{aligned}$$

Algebraic structures of this kind, which are commutative, associative, and which have an identity and inverses, are called *commutative groups*. Thus the complex numbers of unit absolute value form a commutative group under multiplication; rotation mappings of $\mathbf{R}^2$ form a commutative group under composition. We have proved that these two groups are algebraically indistinguishable: given a complex number of unit absolute value, $a+bi$, there corresponds to it a unique rotation mapping, $\rho_{(a,b)}$, and *vice versa*; under this correspondence the product of two such complex numbers is carried into the composition of the associated mappings. We say the two groups are *isomorphic* to one another.

Worked examples 5.2

1 Compute the composition of the rotation mappings, (i) $\rho_{(0,1)}$ and $\rho_{(0,-1)}$, (ii) $\rho_{(1/\sqrt{2},1/\sqrt{2})}$ and $\rho_{(0,1)}$. Interpret your answers geometrically.

(i) Since $(i)(-i) = 1$, we have

$$\rho_{(0,1)} \circ \rho_{(0,-1)} = \rho_{(1,0)}.$$

In geometrical terms, $\rho_{(0,-1)}$ rotates $\mathbf{R}^2$ through an angle $\pi/2$ in a clockwise direction, whereas $\rho_{(0,1)}$ rotates $\mathbf{R}^2$ through an angle $\pi/2$ in an anticlockwise direction. The composition of the two leaves all the vectors in $\mathbf{R}^2$ fixed.

(ii) Since $\left(\frac{1}{\sqrt{2}}+\frac{i}{\sqrt{2}}\right)(i) = -\frac{1}{\sqrt{2}}+\frac{i}{\sqrt{2}}$, we have

$$\rho_{(1/\sqrt{2},\,1/\sqrt{2})} \circ \rho_{(0,1)} = \rho_{(-1/\sqrt{2},\,1/\sqrt{2})}.$$

In geometrical terms, $\rho_{(0,1)}$ rotates $\mathbf{R}^2$ through an angle $\pi/2$ in an anticlockwise direction, and $\rho_{(1/\sqrt{2},\,1/\sqrt{2})}$ through an angle $\pi/4$ in the same direction. Their composition $\rho_{(-1/\sqrt{2},\,1/\sqrt{2})}$ rotates $\mathbf{R}^2$ through an angle $3\pi/4$, anticlockwise, as expected since $\pi/2+\pi/4 = 3\pi/4$.

2 Interpret geometrically the effect of applying the rotation mapping $\rho_{(0,1)}$ to $\mathbf{R}^2$, four times in succession. Describe the corresponding effect in the group of complex numbers of unit absolute value.

The mapping $\rho_{(0,1)}$ rotates $\mathbf{R}^2$ through an angle $\pi/2$. Applying it four times in succession, i.e. applying to $\mathbf{R}^2$ the rotation $\rho^4_{(0,1)}$, will leave all vectors in $\mathbf{R}^2$ fixed, since $4\pi/2 = 2\pi$. Algebraically $\rho_{(0,1)}$ corresponds to multiplication by i; applying $\rho^4_{(0,1)}$ thus involves multiplication by i^4. But $i^4 = (i^2)^2 = (-1)^2 = 1$. Multiplication of complex numbers by 1 leaves them unaltered, just as rotating by $\rho^4_{(0,1)}$ leaves vectors in $\mathbf{R}^2$ unaltered.

3 Find all rotation mappings ρ such that

(i) $\rho^2 = \rho_{(1,0)}$, (ii) $\rho^2 = \rho_{(0,1)}$, (iii) $\rho^3 = \rho_{(1,0)}$.

(i) Suppose $\rho = \rho_{(a,b)}$ and $\rho^2 = \rho_{(1,0)}$. Then $(a+ib)^2 = 1+0i$. Thus $a^2-b^2 = 1$ and $2ab = 0$. If $a = 0$ then $b^2 = -1$, which is not possible since b is real. So we must have $b = 0$ and $a^2 = 1$. Thus $a = \pm 1$, and $\rho = \rho_{(1,0)}$ or $\rho = \rho_{(-1,0)}$.

(ii) Suppose $\rho = \rho_{(a,b)}$ and $\rho^2 = \rho_{(0,1)}$. Then $(a+ib)^2 = 0+i = i$. Thus $a^2-b^2 = 0$ and $2ab = 1$. Hence $a = \pm b$ and $2a^2 = 1$; i.e. $a = \pm 1/\sqrt{2}$ and $b = \pm 1/\sqrt{2}$, the $\pm$ signs being taken together. In complex number terms $a+ib = \pm(1/\sqrt{2})(1+i)$. Thus $\rho = \rho_{(1/\sqrt{2},\, 1/\sqrt{2})}$ or $\rho = \rho_{(-1/\sqrt{2},\, -1/\sqrt{2})}$.

(iii) Suppose $\rho = \rho_{(a,b)}$ and $\rho^3 = \rho_{(1,0)}$. Then $(a+ib)^3 = 1+0i$. Setting $z = a+ib$, $z^3 = 1$ so that

$$(z-1)(z^2+z+1) = 0.$$

Thus $z = 1$ or $\frac{1}{2}(-1 \pm i\sqrt{3})$. In rotational terms, $\rho = \rho_{(1,0)}$ or $\rho = \rho_{(-1/2,\, -\sqrt{3}/2)}$ or $\rho = \rho_{(-1/2,\, -\sqrt{3}/2)}$.

Exercises 5.2

1 Prove algebraically that the composition of clockwise rotations through $\pi/2$ and $3\pi/2$ leaves vectors in $\mathbf{R}^2$ fixed. Describe the corresponding effect in the group of complex numbers of unit absolute value.

2 Find a rotation mapping $\rho_{(a,b)} \neq \rho_{(1,0)}$ which when applied to $\mathbf{R}^2$ six times in succession leaves all vectors $\mathbf{R}^2$ fixed.

3 In the light of your answer to Question 2 above, can you find a complex number z such that $z^6 = 1$? Is your answer unique?

5.3 Measurement of rotation mappings of $\mathbf{R}^2$

We shall now introduce into $\mathbf{R}^2$ certain basic trigonometrical ideas which will eventually be sufficient to justify us formally transferring the traditional techniques of elementary trigonometry, which we have already been using informally, from the real plane into the vector space $\mathbf{R}^2$.

Our definition of a rotation mapping of $\mathbf{R}^2$ in section 5.1 already began this process of transfer, since the algebraic definition which we gave there in $\mathbf{R}^2$ corresponds precisely to the first trigonometrical step we make in studying 'angles' in the real plane, when we consider them as an outcome of our understanding of the idea of 'turning about a point'. The next step in our elementary trigonometrical study would lead us to measure amounts of turning and involve the use of a 'protractor', in some form or other, to measure angles in degrees or radians. Thus, if we are to continue the transfer to $\mathbf{R}^2$ of our pattern of development in elementary geometry and trigonometry, we next require in $\mathbf{R}^2$ an 'algebraic protractor', which will enable us to measure our rotation mappings of $\mathbf{R}^2$.

In elementary geometry we are given a protractor and accept it as a means of measurement in the same way as we accept the use of a ruler in linear measurement. Similarly here we shall state without proof a theorem asserting the existence of adequate algebraic structure associated with $\mathbf{R}^2$. (A proper mathematical justification of our assertions would take us beyond the scope of a book such as this, and would require a proof of a non-algebraic, analytical nature.)

Our algebraic protractor, like the one we use in elementary geometry, is designed to give us a linear measurement around the circle in terms of real numbers. We require in effect to 'map' the real line linearly around the circle. To do this we shall associate with our rotation mappings the circle, centre the origin and of unit radius, around which the 'ends' of our vectors of unit length are carried by our mappings. We then relate each real number x to a point P_x on the circle and hence to a rotation which carries the point $(1, 0)$ to P_x, or equivalently to the complex number of unit absolute value which is represented by P_x. Of course, many real numbers will correspond to the same point. Such real numbers will differ by integer multiples of 2π. Pictorially we can illustrate what we require to happen as shown overleaf (in which the upper half circle is the protractor of our elementary trigonometry).

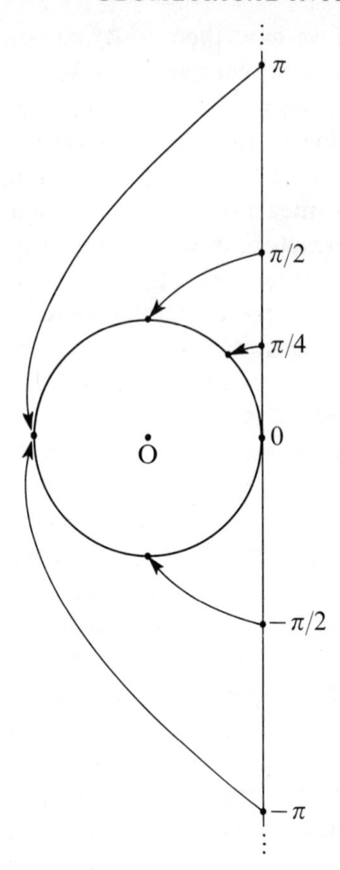

Specifically, then, we insist on the truth of the following result.

Protractor theorem. (i) *To every real number x there corresponds a rotation mapping* θ_x *of* $\mathbf{R}^2$, *and to every rotation mapping* θ *there corresponds an infinite set of real numbers* x *such that* $\theta = \theta_x$ *for all* x *in the set.*

(ii) *There exists a real number* π *such that two real numbers* x *and* y *correspond to the same rotation mapping, i.e.* $\theta_x = \theta_y$, *if and only if* $x - y = 2n\pi$ *for some integer* n.

(iii) *If* $0 \leqslant x \leqslant \pi$ *and* $\theta_x = \rho_{(a,b)}$, *then* $b \geqslant 0$.

This correspondence between sets of real numbers differing from each other by integer multiples of 2π and rotation mappings of **R**^{2} gives us our required (angular) measurement of these mappings. Any of the real numbers $x+2n\pi$ is said to measure θ_x and we now say θ_x is a rotation of **R**^{2} through x radians, or *through an angle* x†.

Since any natural measuring process must have the property that we can add or subtract measurements in a satisfactory fashion in relation to the objects we are measuring, we must make a further statement to this effect. To do this we use the group structure of the rotation mappings of **R**^{2} which we introduced in the last section, and add to our theorem as follows.

(iv) *If real numbers x and y correspond to rotation mappings θ_x and θ_y respectively, then the sum $x+y$ of x and y corresponds to the composition $\theta_x \circ \theta_y$ of θ_x and θ_y: the difference $x-y$ of x and y corresponds to the composition $\theta_x \circ \theta_y^{-1}$ of θ_x and the inverse θ_y^{-1} of θ_y: the integer multiples $2n\pi$ correspond to the identity rotation $\rho_{(1,0)}$.*

With this extended version of our theorem we can formally calculate in a relatively simple but now rigorous algebraic fashion, in **R**^{2}, the rotation mappings through angles such as π, $\pi/2$, $\pi/3$, $\pi/4$. Consider, for example, θ_π and $\theta_{\pi/2}$ as illustrations of the kind of formal argument in **R**^{2} which can now replace the informal geometrical approach we have previously used in the plane. Of course we must get the answers we expect in the light of our informal arguments, namely that a rotation through π carries $(1, 0)$ to $(-1, 0)$, so that $\theta_\pi = \rho_{(-1,0)}$, and a rotation through $\pi/2$ carries $(1, 0)$ to $(0, 1)$, i.e. $\theta_{\pi l 2} = \rho_{(0,1)}$.

Consider θ_π first. Since $2\pi = \pi+\pi$, we have from part (iv) of our Protractor theorem, $\theta_{2n} = \theta_\pi \circ \theta_x = \theta_\pi^2$. But $\theta_{2\pi} = \rho_{(1,0)}$, so that if $\theta_\pi = \rho$ we require to solve the equation

$$\rho^2 = \rho_{(1,0)}.$$

This we have already done in Worked example 3 of the last

†Only in this last sense will we make formal use of the word 'angle'. In so doing we are in effect following standard elementary practice, identifying 'angles' with their measurement.

section, where we found $\rho = \rho_{(1,0)}$ or $\rho_{(-1,0)}$. Here we cannot have $\theta_\pi = \rho_{(1,0)} = \theta_{2\pi}$, since if $\theta_x = \theta_y$, x and y must differ by an integer multiple of 2π, and $2\pi - \pi = \pi$ is not such a multiple. Thus $\rho_{(-1,0)}$ is the required solution and as we expected, $\theta_\pi = \rho_{(-1,0)}$.

Now consider $\theta_{\pi/2}$. Since $\pi = \pi/2 + \pi/2$, we have $\theta_\pi = \theta_{\pi/2} \circ \theta_{\pi/2} = \theta_{\pi/2}^2$. But $\theta_\pi = \rho_{(-1,0)}$, so that if $\theta_\pi = \rho$ we have to solve the equation

$$\rho^2 = \rho_{(-1,0)}.$$

Once again we set $\rho = \rho_{(a,b)}$, in which case $(a+ib)^2 = -1+0i$. Thus $a^2 - b^2 = -1$ and $2ab = 0$. It follows that either $b = 0$ and $a^2 = -1$, which is not possible since a is real, or $a = 0$ and $b = \pm 1$. Since $0 < \pi/2 < \pi$, part (iii) of our theorem implies $b = +1$. Thus $\theta_\pi = \rho_{(0,1)}$, again as expected.

Algebraically we now have a situation in which the additive structure of the real numbers is mirrored in the algebraic theory of rotation mappings, or equivalently in the algebra of complex numbers of unit absolute value. On the one hand we have a commutative group consisting of the real numbers under addition; the sum of any two real numbers in any order is a real number; this addition is associative; there is an identity, namely 0, which when added to any real number leaves it unaltered; finally any real number has associated with it its negative, which when added to it gives 0. On the other hand we have the commutative group of rotation mappings of $\mathbf{R}^2$, or of the complex numbers of unit absolute value. The correspondence $x \to \theta_x$ between the (additive) group of real numbers and the group of rotation mappings carries sums of real numbers $x+y$ to the composition of mappings $\theta_x \circ \theta_y$ and so 'preserves structure', even though it is not a one-one correspondence, carrying, as it does, an infinity of real numbers to any given rotation.

Worked examples 5.3

1 Prove $\theta_{\pi/3} = \rho_{(1/2, \sqrt{3}/2)}$.

Let $\theta_{\pi/3} = \rho_{(a,b)} = \rho$. Then since $\theta_{\pi/3} \circ \theta_{\pi/3} \circ \theta_{\pi/3} = \theta_\pi$, we have $\rho^3 = \rho_{(-1,0)}$. But

$$(a+ib)^3 = (a^3-3ab^2)+i(3a^2b-b^3).$$

Thus $(a+ib)^3 = -1+0i$ (which follows from our equation $\rho^3 = \rho_{(-1,0)}$) implies

$$a^3-3ab^2 = -1,$$

and

$$3a^2b-b^3 = 0.$$

The second of these equations implies $b = 0$ or $3a^2 = b^2$. If $b = 0$, $a^3 = -1$, whence $a = -1$ since a is real. But then $\theta_{\pi/3} = \rho_{(-1,0)} = \theta_\pi$ and this is not possible since $\pi-\pi/3 = 2\pi/3$ which is not an integer multiple of 2π. Thus $3a^2 = b^2$ and the first equation implies $8a^3 = 1$ which, again since a is real, implies $a = \frac{1}{2}$. But then $b = \pm\sqrt{3}/2$ and since $0 < \pi/3 < \pi$, $b = +\sqrt{3}/2$. Thus $\theta_{\pi/3} = \rho_{(1/2,\sqrt{3}/2)}$, as required.

2 Prove $\theta_{-\pi/2} = \rho_{(0,-1)}$.

The rotation $\theta_{-\pi/2}$ is 'inverse' to $\theta_{\pi/2} = \rho_{(0,1)}$, so that if $\theta_{-\pi/2} = \rho_{(a,b)}$, we have

$$\rho_{(a,b)} \circ \rho_{(0,1)} = \rho_{(1,0)}$$

i.e.

$$\rho_{(-b,a)} = \rho_{(1,0)}.$$

So $b = -1$ and $a = 0$, whence $\theta_{-\pi/2} = \rho_{(0,-1)}$, as required.

Exercises 5.3

1 Prove $\theta_{2\pi/3} = \rho_{(-1/2,\pi 3/2)}$.
(Hint: $\theta_{2\pi/3} \circ \theta_{\pi/3} = \theta_\pi$).

2 Prove $\theta_{\pi/4} = \rho_{(1/\sqrt{2},1/\sqrt{2})}$.
(Hint: $\pi/4+\pi/4 = \pi/2$).

3 Prove $\theta_{-\pi/4} = \rho_{(1/\sqrt{2},-1/\sqrt{2})}$.

4 Use the Protractor theorem to compute rotation mappings $\rho_{(a,b)}$ corresponding to the real numbers $3\pi/2$, $-\pi/2$, $-\pi/3$, $3\pi/4$.

5.4 Trigonometry in $\mathbf{R}^2$

We now finally formalize the use of elementary trigonometry in $\mathbf{R}^2$ by defining sines and cosines of angles by means of rotation mappings. We take our lead from elementary trigonometry, and start in effect from the definitions 'sine equals opposite over hypotenuse', 'cosine equals adjacent over hypotenuse', together with their extensions to angles greater than $\pi/2$. The real numbers a, b such that $a^2+b^2 = 1$, which are associated with a rotation $\rho_{(a,b)}$, give us the required 'adjacent' and 'opposite' sides of a right angled triangle with hypotenuse of unit length.

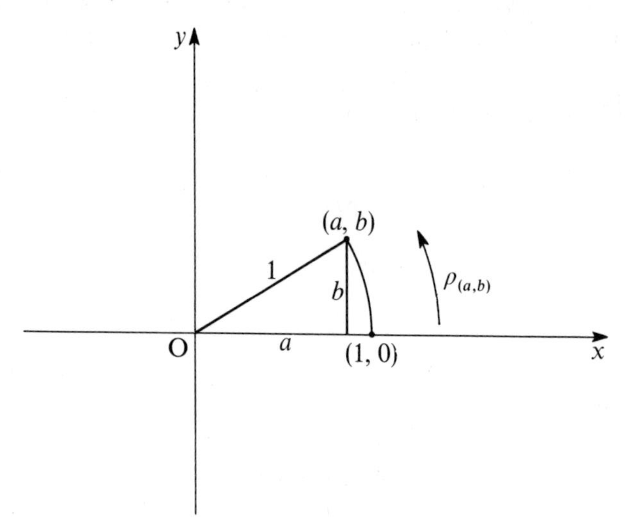

We therefore begin by defining the cosine and sine of a rotation mapping of $\mathbf{R}^2$ by the formulae

$$\text{cosine}(\rho_{(a,b)}) = a, \quad \text{sine}(\rho_{(a,b)}) = b.$$

These formulae are a sufficient basis for a trigonometrical theory of rotation mappings in the vector space $\mathbf{R}^2$; however it is naturally our wish to work in the more familiar context of angles in the plane rather than rotations. To produce this more familiar situation we use the correspondence $x \to \theta_x$ with which we measured our rotation mappings. We therefore define

$$\text{cosine}(x) = \text{cosine}(\theta_x)$$

$$\text{sine}(x) = \text{sine}(\theta_x).$$

In conventional fashion we abbreviate cosine (x), writing it $\cos x$, and sine (x), writing it $\sin x$.

In the worked examples which now follow we carry out typical computations deriving well-known elementary trigonometrical formulae. In each case, of necessity, because of our definitions, we fall back on cosines and sines of rotation mappings. This we must do if we are to base our work formally on the algebraic structure of $\mathbf{R}^2$. The examples and exercises are intended to convince the reader that the basic structure of rotation mappings of $\mathbf{R}^2$ was well chosen to lead us quickly and surely back to the familiar ground of elementary trigonometry, but now with a precision given to us by our study of the vector space $\mathbf{R}^2$, rather than the imprecision which of necessity we accept in an initial elementary trigonometrical study of angles and their measurement in the real plane.

Worked examples 5.4

1 Prove $\cos \pi/2 = 0$, $\sin \pi/2 = 1$, $\cos \pi/4 = 1/\sqrt{2}$, $\sin \pi/4 = 1/\sqrt{2}$.

As suggested above, in computations such as these we simply refer back to the appropriate rotation mapping $\rho_{(a,b)}$ and read off a and b. Thus the rotation mapping through $\pi/2$ is $\theta_{\pi/2}$ which we have already computed to be $\rho_{(0,1)}$. Thus $\cos \pi/2 = 0$ and $\sin \pi/2 = 1$.

The rotation mapping corresponding to $\pi/4$ is $\theta_{\pi/4}$ which Exercise 2 of Exercises 5.3 asserts is $\rho_{(1/\sqrt{2},\,1/\sqrt{2})}$.

Thus, given the result of that exercise, $\cos \pi/4 = \sin \pi/4 = 1/\sqrt{2}$, as required.

2 Deduce from first principles in $\mathbf{R}^2$, the cosine formulae

$$\cos(x+2n\pi) = \cos x, \cos 2n\pi = 1, \cos(-x) = x,$$

and

$$\cos(x+y) = \cos x \cos y - \sin x \sin y;$$

and the corresponding sine formulae

$$\sin(x+2n\pi) = \sin x, \sin 2n\pi = 0, \sin(-x) = -\sin x,$$

and

$$\sin(x+y) = \cos x \sin y + \sin x \cos y.$$

The first are trivial, since by definition x and $x+2n\pi$ measure the same rotation mapping θ_x, so that

$$\cos x = \text{cosine}(\theta_x) = \cos(x+2n\pi),$$

and

$$\sin x = \text{sine}(\theta_x) \quad = \sin(x+2n\pi).$$

Since $2n\pi$ measures the identity mapping $\rho_{(1,0)}$, we have

$$\cos 2n\pi = \text{cosine}(\rho_{(1,0)}) = 1$$

$$\sin 2n\pi = \text{sine}(\rho_{(1,0)}) \quad = 0.$$

If x measures a rotation mapping $\rho_{(a,b)}$, $-x$ must measure the inverse $\rho^{-1}_{(a,b)}$. But we have seen that $\rho^{-1}_{(a,b)} = \rho_{(a,-b)}$, so that

$$\begin{aligned}\cos(-x) &= \text{cosine}(\rho^{-1}_{(a,b)})\\ &= \text{cosine}(\rho_{(a,-b)})\\ &= a\\ &= \text{cosine}(\rho_{(a,b)})\\ &= \cos x,\end{aligned}$$

and similarly

$$\begin{aligned}
\sin(-x) &= \text{sine}(\rho_{(a,b)}^{-1}) \\
&= \text{sine}(\rho_{(a,-b)}) \\
&= -b \\
&= -\text{sine}(\rho_{(a,b)}) \\
&= -\sin x.
\end{aligned}$$

Finally, the derivation of the addition formulae is just as straightforward. Suppose x measures the rotation mapping $\rho_{(a,b)}$, and y measures $\rho_{(c,d)}$. Then $x+y$ measures the composition

$$\rho_{(a,b)} \circ \rho_{(c,d)} = \rho_{(ac-bd,\, ad+bc)}.$$

Thus

$$\begin{aligned}
\cos(x+y) &= \text{cosine}(\rho_{(ac-bd,\, ad+bc)}) \\
&= ac-bd \\
&= \text{cosine}(\rho_{(a,b)})\,\text{cosine}(\rho_{(c,d)}) \\
&\qquad - \text{sine}(\rho_{(a,b)})\,\text{sine}(\rho_{(c,d)}) \\
&= \cos x \cos y - \sin x \sin y,
\end{aligned}$$

and similarly

$$\begin{aligned}
\sin(x+y) &= \text{sine}(\rho_{(ac-bd,\, ad+bc)}) \\
&= ad+bc \\
&= \cos x \sin y + \sin x \cos y.
\end{aligned}$$

3 Prove that if ρ is any rotation mapping of $\mathbf{R}^2$, then

$$\text{cosine}(\rho \circ \rho) = 2(\text{cosine}(\rho))^2 - 1.$$

Suppose $\rho = \rho_{(a,b)}$, where as usual $a^2+b^2 = 1$. Then

$$\begin{aligned}
\rho \circ \rho &= \rho_{(a,b)} \circ \rho_{(a,b)} \\
&= \rho_{(a^2-b^2,\, 2ab)}
\end{aligned}$$

Thus $$\begin{aligned}\text{cosine}(\rho \circ \rho) &= a^2 - b^2\\ &= 2a^2 - 1 \quad (\text{since } a^2 + b^2 = 1)\\ &= 2(\text{cosine}(\rho))^2 - 1.\end{aligned}$$

Exercises 5.4

1 Prove $\cos \pi/3 = \frac{1}{2}$, $\sin \pi/3 = \sqrt{3}/2$, $\cos 2\pi/3 = -\frac{1}{2}$, $\sin 2\pi/3 = \sqrt{3}/2$.

2 Deduce from the formula for $\cos(x+y)$, that $\cos(x+\pi) = -\cos x$.

3 Prove from first principles in $\mathbf{R}^2$ that

$$1 = \cos^2 x + \sin^2 x,$$

and $$\sin 2x = 2 \sin x \cos x.$$

4 Prove that, if ρ is any rotation mapping of $\mathbf{R}^2$, then

$$\text{cosine}(\rho \circ \rho \circ \rho) = 4(\text{cosine}(\rho))^3 - 3\,\text{cosine}(\rho).$$

Deduce that

$$\cos 3x = 4\cos^3 x - 3\cos x,$$

for any real number x.

5.5 Complex numbers in polar form

Given any non-zero complex number $z = x + yi$, if we set $r = \sqrt{(x^2+y^2)}$, then $r \neq 0$ and we can write

$$z = r\left(\frac{x}{r} + \frac{y}{r}i\right).$$

Since the complex number $\frac{x}{r} + \frac{y}{r}i$ has unit absolute value, it defines a rotation mapping $\rho_{(x/r,\, y/r)}$, through some angle θ such that

$$\cos\theta = \frac{x}{r}, \quad \sin\theta = \frac{y}{r}.$$

We can thus write z 'in polar form':

$$z = r(\cos\theta + i\sin\theta).$$

Of course it does not matter which real number in the set $\{\theta+2n\pi\}$ we choose to measure $\rho_{(x/r,\, y/r)}$. However it is usual in this context to pick out one of the real numbers $\theta+2n\pi$ called the *principal value of the argument of* z and defined as follows.

Definition. Given any complex number $x+yi$ such that $r = \sqrt{(x^2+y^2)}$, the principal value of the argument of $x+yi$ is the real number denoted by $\arg(x+yi)$ which measures the rotation mapping $\rho_{(x/r,\, y/r)}$ and is such that $-\pi < \arg(x+yi) \leqslant \pi$.

The phrase 'argument of $x+yi$' as opposed to 'principal value of the argument of $x+yi$' is reserved to denote the whole set of numbers $\{\arg(x+yi)+2n\pi\}$, for all integers n.

Pictorially we have the traditional polar co-ordinate form of the real plane.

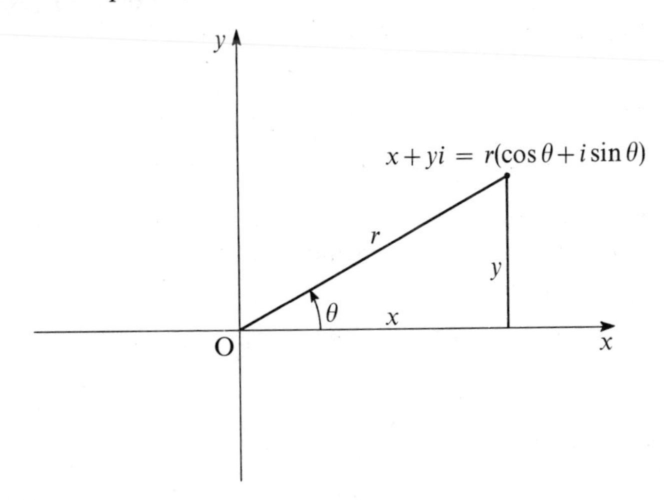

In polar form, multiplication of complex numbers completely reflects our interpretation of multiplication by means of rotations, for our elementary trigonometry of $\mathbf{R}^2$ yields

$$\begin{aligned}(\cos\theta + i\sin\theta)(\cos\phi + i\sin\phi) &= (\cos\theta\cos\phi - \sin\theta\sin\phi) \\ &\quad + i(\cos\theta\sin\phi + \sin\theta\cos\phi) \\ &= \cos(\theta+\phi) + i\sin(\theta+\phi).\end{aligned}$$

In other words, as we would expect, 'turning through an angle θ' followed by 'turning through an angle ϕ', surely 'turns the plane through an angle $\theta+\phi$'. (Notice however that care should be taken in this context if principal values are required: if θ and ϕ are principal values it does not follow that $\theta+\phi$ will be a principle value, i.e. $-\pi < \theta$, $\phi \leqslant \pi$ does not imply $-\pi < \theta+\phi \leqslant \pi$,)

If in the above multiplication we set $\theta = \phi$ we have

$$(\cos\theta + i\sin\theta)^2 = \cos 2\theta + i\sin 2\theta;$$

i.e. two rotations through an angle θ are equivalent to a rotation through 2θ. Of course such a formula generalizes, and one can confirm formally as follows what is clear in intuitive rotational terms, namely that m rotations through an angle θ produce a rotation through $m\theta$.

De Moivre's theorem. *If m is any positive integer*

$$(\cos\theta + i\sin\theta)^m = \cos m\theta + i\sin m\theta.$$

Proof. We proceed by induction on m. The result is obviously true for $m = 1$. Suppose therefore it is true for $m = r-1$, so that

$$(\cos\theta + i\sin\theta)^{r-1} = \cos(r-1)\theta + i\sin(r-1)\theta.$$

Then

$$\begin{aligned}(\cos\theta + i\sin\theta)^r &= (\cos\theta + i\sin\theta)(\cos\theta + i\sin\theta)^{r-1} \\ &= (\cos\theta + i\sin\theta)(\cos(r-1)\theta + i\sin(r-1)\theta) \\ &= \cos(\theta + (r-1)\theta) + i\sin(\theta + (r-1)\theta) \\ &= \cos r\theta + i\sin r\theta,\end{aligned}$$

and the required result follows by induction on m.

The result of de Moivre's theorem holds true even when m is a non-positive integer. In case $m = 0$ the proof is a formality: in defining the integral powers z^m of a complex number we have $z^0 = 1$ *by definition*, so that

$$\begin{aligned}(\cos\theta+i\sin\theta)^0 &= 1\\ &= \cos 0+i\sin 0,\end{aligned}$$

since $\cos 2n\pi = 1$, $\sin 2n\pi = 0$ for any integer n. In case m is negative, say $m = -n$, we again have by definition, $z^m = z^{-n} = (z^n)^{-1}$. Thus

$$(\cos\theta+i\sin\theta)^m = ((\cos\theta+i\sin\theta)^n)^{-1}.$$

But $$(\cos\theta+i\sin\theta)^n = \cos n\theta+i\sin n\theta,$$

and $$\begin{aligned}(\cos n\theta+i\sin n\theta)^{-1} &= \cos n\theta-i\sin n\theta\\ &= \cos(-n\theta)+i\sin(-n\theta),\end{aligned}$$

by our previous computations. Thus,

$$(\cos\theta+i\sin\theta)^m = \cos m\theta+i\sin m\theta,$$

even if m is a negative integer.

From now on we shall make increasing use of possibly more complicated trigonometrical formulae in $\mathbf{R}^2$. We shall therefore assume the reader is convinced that any results from elementary trigonometry can be proved in $\mathbf{R}^2$, and we shall assume any such results, as necessary.

Worked examples 5.5

1 Write in polar form the complex number $1+i$.

The absolute value of $1+i$ is $\sqrt{2}$. We have

$$1+i = \sqrt{2}\left(\frac{1}{\sqrt{2}}+\frac{i}{\sqrt{2}}\right),$$

whence, in polar form,

$$1+i = \sqrt{2}\left(\cos\frac{\pi}{4}+i\sin\frac{\pi}{4}\right).$$

2 Find the argument, and the principal value of the argument, of the complex number $1+\sin\alpha+i\cos\alpha$ if $-\pi < \alpha < -\pi/2$.

By elementary trigonometrical arguments (!) we have

$$\begin{aligned}
1+\sin\alpha+i\cos\alpha &= 1+\cos\left(\frac{\pi}{2}-\alpha\right)+i\sin\left(\frac{\pi}{2}-\alpha\right) \\
&= 2\cos^2\left(\frac{\pi}{4}-\frac{\alpha}{2}\right)+2i\sin\left(\frac{\pi}{4}-\frac{\alpha}{2}\right)\cos\left(\frac{\pi}{4}-\frac{\alpha}{2}\right) \\
&= 2\cos\left(\frac{\pi}{4}-\frac{\alpha}{2}\right)\left(\cos\left(\frac{\pi}{4}-\frac{\alpha}{2}\right)+i\sin\left(\frac{\pi}{4}-\frac{\alpha}{2}\right)\right).
\end{aligned}$$

Now since $-\pi < \alpha < -\frac{\pi}{2}$, it follows that $\frac{\pi}{2} < \frac{\pi}{4}-\frac{\alpha}{2} < \frac{3\pi}{4}$.

Thus $2\cos\left(\frac{\pi}{4}-\frac{\alpha}{2}\right) < 0$ and the absolute value of $1+\sin\alpha+i\cos\alpha$ is in this case $-2\cos\left(\frac{\pi}{4}-\frac{\alpha}{2}\right)$. Finally, then,

$$\begin{aligned}
&1+\sin\alpha+i\cos\alpha \\
&= \left(-2\cos\left(\frac{\pi}{4}-\frac{\alpha}{2}\right)\right)\left(-\cos\left(\frac{\pi}{4}-\frac{\alpha}{2}\right)-i\sin\left(\frac{\pi}{4}-\frac{\alpha}{2}\right)\right) \\
&= \left(-2\cos\left(\frac{\pi}{4}-\frac{\alpha}{2}\right)\right)\left(\cos\left(\frac{5\pi}{4}-\frac{\alpha}{2}\right)+i\sin\left(\frac{5\pi}{4}-\frac{\alpha}{2}\right)\right)
\end{aligned}$$

since $\cos(\pi+x) = -\cos x$ and $\sin(\pi+x) = -\sin x$. The required argument is therefore $\frac{5\pi}{4}-\frac{\alpha}{2}+2n\pi$; the principal value is given by taking $n = -1$, and is $\frac{5\pi}{4}-\frac{\alpha}{2}-2\pi = -\left(\frac{3\pi}{4}+\frac{\alpha}{2}\right)$.

3 Use the binomial theorem on the one hand and de Moivre's theorem on the other, to evaluate $(\cos\theta + i\sin\theta)^5$. Deduce that

$$\sin 5\theta = 16\sin^5\theta - 20\sin^3\theta + 5\sin\theta.$$

By the binomial theorem,

$$\begin{aligned}(\cos\theta + i\sin\theta)^5 = {} & \cos^5\theta + 5i\cos^4\theta\sin\theta \\ & -10\cos^3\theta\sin^2\theta \\ & -10i\cos^2\theta\sin^3\theta \\ & +5\cos\theta\sin^4\theta + i\sin^5\theta.\end{aligned}$$

By de Moivre's theorem,

$$(\cos\theta + i\sin\theta)^5 = \cos 5\theta + i\sin 5\theta.$$

Thus

$$\begin{aligned}\sin 5\theta &= 5\cos^4\theta\sin\theta - 10\cos^2\theta\sin^3\theta + \sin^5\theta \\ &= 5(1-\sin^2\theta)^2\sin\theta - 10(1-\sin^2\theta)\sin^3\theta + \sin^5\theta \\ &= 16\sin^5\theta - 20\sin^3\theta + 5\sin\theta.\end{aligned}$$

Exercises 5.5

1 Write the following complex numbers in polar form: (i) 1, (ii) i, (iii) $1-i$, (iv) $\sqrt{3}-i$, (v) $(1-i)^2$.

2 Find the principal values of the arguments of the complex numbers in Question 1.

3 Compare the principal value of the complex numbers i and -1 with that of their product $-i$.

4 Find the absolute value and argument of

$$\left(1 + \sin\frac{\pi}{3} + i\cos\frac{\pi}{3}\right)^5.$$

5 If $z = \cos\theta + i\sin\theta$, prove that $2\cos\theta = z + 1/z$, $2i\sin\theta = z - 1/z$.

6 If $z = \cos\theta + i\sin\theta$, express $z^n + 1/z^n$ and $z^n - 1/z^n$ in terms of $\cos n\theta$ and $\sin n\theta$.

7 Use the results of Questions 5 and 6 to prove that

$$16\cos^5\theta = \cos 5\theta + 5\cos 3\theta + 10\cos\theta.$$

8 Find the absolute value and argument of $(1+i)^m$ and of $(\sqrt{3}-i)^m$.

5.6 Complex roots of unity

De Moivre's theorem has given us a simple technique for calculating integral powers of complex numbers. We are left therefore with the problem of calculating rational powers in **C**, i.e. with the problem of finding roots of complex numbers.

In the real field **R** we recall that given a real number a, an equation $x^m = a$ has one real (negative) solution if m is odd and $a < 0$. It has no solutions if m is even and $a < 0$. Finally it has one solution (positive) or two solutions (one positive, the other negative) according as m is odd or even, if $a > 0$. We reserve the notation $\sqrt[m]{a}$ for the (unique) positive real mth root of $x^m = a$, in case $a > 0$.

In the complex field we find a more complicated situation. An equation such as $z^m = w$ is of degree m in z and since as we saw on page 129, **C** is algebraically closed, we must expect m complex solutions (which may or may not be distinct).

Consider a simple equation, $z^3 = 1$. Obviously $z = 1$ is a solution. But de Moivre's theorem implies that

$$z = \cos\frac{2\pi}{3} + i\sin\frac{2\pi}{3}$$

is also a solution, since

$$\left(\cos\frac{2\pi}{3} + i\sin\frac{2\pi}{3}\right)^3 = \cos 2\pi + i\sin 2\pi$$

$$= 1.$$

Again $z = \cos\frac{4\pi}{3}+i\sin\frac{4\pi}{3}$ is also a solution, as is $z = \cos\left(-\frac{2\pi}{3}\right)+i\sin\left(-\frac{2\pi}{3}\right)$, and so on. In fact $z = \cos\frac{2n\pi}{3}+i\sin\frac{2n\pi}{3}$ for any integer n is a solution. However these numbers are not all distinct. We can pick out three distinct solutions by taking $n = 0$, 1 and 2, when we have $z = 1$, $-\frac{\sqrt{3}}{2}+\frac{1}{2}i$ or $-\frac{\sqrt{3}}{2}-\frac{1}{2}i$, each of which has equal right to be called a cube root of unity. In the real field $(1)^{1/3}$ is uniquely defined as 1; in the complex field this is no longer true.

Geometrically, in terms of rotations, we see the same problem. A rotation through $2\pi/3$ is not the only rotation which when repeated three times is equivalent to the identity rotation: there is more than one solution ρ of the equation $\rho^3 = \rho_{(1,0)}$ in the group of rotation mappings of $\mathbf{R}^2$. But only three such rotation mappings are distinct, namely the identity rotation, the rotation through $2\pi/3$ and the rotation through $4\pi/3$.

In general, suppose we are given in $\mathbf{C}$ an equation $z^m = w$, where $|w| = r$. Then since $|zw| = |z||w|$ for any z, w, we have

$$|z|^m = |z^m| = |w| = r,$$

so that if z is such that $z^m = w$, it follows that $|z|$, being a positive real number by definition, is equal to the positive mth root of $|w|$, i.e. $|z| = \sqrt[m]{r}$.

Let $s = \sqrt[m]{r}$, and in polar form let

$$w = r(\cos\alpha+i\sin\alpha),$$

and

$$z = s(\cos\theta+i\sin\theta).$$

Our equation now takes the form

$$(\cos\theta+i\sin\theta)^m = \cos\alpha+i\sin\alpha.$$

By de Moivre's theorem we have solutions

$$z = \cos\left(\frac{\alpha+2k\pi}{m}\right)+i\sin\left(\frac{\alpha+2k\pi}{m}\right)$$

for any integer value of k, since

$$\left(\cos\left(\frac{\alpha+2k\pi}{m}\right)+i\sin\left(\frac{\alpha+2k\pi}{m}\right)\right)^m = \cos(\alpha+2k\pi)+i\sin(\alpha+2k\pi)$$

$$= \cos\alpha+i\sin\alpha.$$

However, as in the particular case $z^3 = 1$ treated above, not all these solutions are distinct, since for any integer n,

$$\begin{aligned}\cos\left(\frac{\alpha+2k\pi}{m}\right)+i\sin\left(\frac{\alpha+2k\pi}{m}\right) &= \cos\left(\frac{\alpha+2k\pi}{m}+2n\pi\right)\\ &\quad +i\sin\left(\frac{\alpha+2k\pi}{m}+2n\pi\right)\\ &= \cos\left(\frac{\alpha+2(k+mn)\pi}{m}\right)\\ &\quad +i\sin\left(\frac{\alpha+2(k+mn)\pi}{m}\right).\end{aligned}$$

We can again, however, select m distinct solutions (which from the general theory of equations in $\mathbf{C}$ constitute all the solutions), by choosing particular values of k. Thus

$$\left\{\cos\left(\frac{\alpha+2k\pi}{m}\right)+i\sin\left(\frac{\alpha+2k\pi}{m}\right)\right\}$$

where $k = 0, 1, \ldots, m-1$ forms such a set of solutions, which are distinct since they correspond to rotations through $\frac{\alpha}{m}, \frac{\alpha+2\pi}{m}, \ldots, \frac{\alpha+2(m-1)\pi}{m}$, no pair of which differs by an integer multiple of 2π.

Collecting this work together we have the following result.

Theorem. *In the complex field* $\mathbf{C}$, *the equation*

$$z^m = w$$

where in polar form $w = r(\cos\alpha+i\sin\alpha)$, *has precisely* m *solutions given by*

$$z = \sqrt[m]{r}\left(\cos\left(\frac{\alpha+2k\pi}{m}\right)+i\sin\left(\frac{\alpha+2k\pi}{m}\right)\right)$$

for $k = 0, 1, \ldots, m-1$, *where* $\sqrt[m]{r}$ *denotes the positive real* mth *root of* $r = |w|$.

In case $|w| = 1$ and $n = 2$, 3 or 4 we have the following pictures.

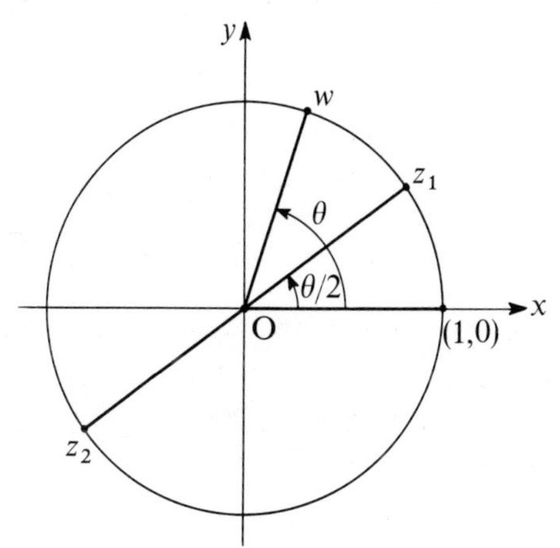

Roots z_1, z_2 of $z^2 = w$

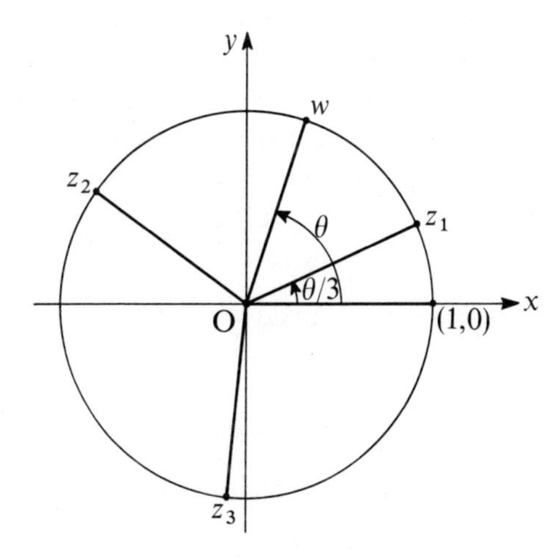

Roots z_1, z_2, z_3 of $z^3 = w$

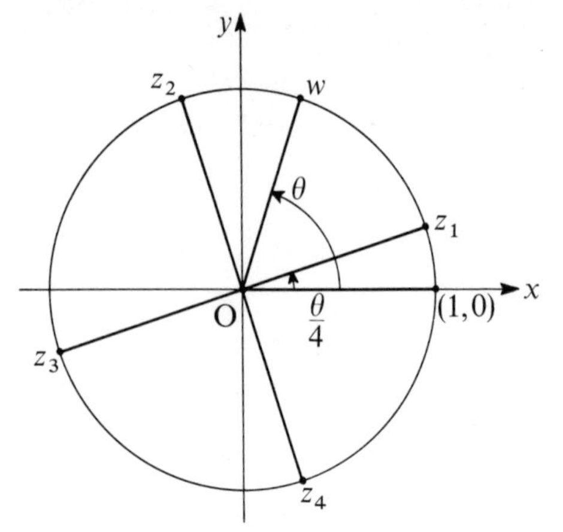

Roots z_1, z_2, z_3, z_4 of $z^4 = \mathrm{w}$

It is usual at this point in the theory of complex numbers to introduce exponential notation, giving a shorthand notation for which there are good analytic reasons. Here we shall write, purely formally,

$$\cos\theta + i\sin\theta = \mathrm{e}^{i\theta}.$$

This notation can be proved to extend the theory of the real exponential function e^x, but since such an analytic theory lies beyond the scope of a book such as this we shall not pursue this point. We reiterate that our definition is a formal shorthand only and we note that in these terms the ordinary real theory of indices goes through, except of course any theory involving finding unique roots, since we have seen there is no single value we can assign to $(\cos\theta + i\ \sin\theta)^{1|m}$ Thus, for example,

$$(\cos\theta + i\sin\theta)(\cos\phi + i\sin\phi) = \cos(\theta+\phi) + i\sin(\theta+\phi)$$

can now be written

$$(\mathrm{e}^{i\theta})(\mathrm{e}^{i\phi}) = \mathrm{e}^{i(\theta+\phi)},$$

or again, de Moivre's theorem implies

$$(\mathrm{e}^{i\theta})^m = \mathrm{e}^{im\theta},$$

for any integer m.

In these terms the mth roots of unity, i.e. of $1 = \cos 2\pi + i \sin 2\pi$, can be conveniently written $\{e^{i(2k\pi/m)}\}$, where $k = 0, 1, \ldots, m-1$.

Worked examples 5.6

1 Solve the equation $z^5 + 1 = 0$ in the complex field. Display your solutions pictorially.

We have to solve

$$\begin{aligned} z^5 &= -1 \\ &= \cos \pi + i \sin \pi \\ &= e^{i\pi}. \end{aligned}$$

Thus we have the five solutions

$$e^{i\pi/5}, \quad e^{i(\pi+2\pi)/5}, \quad e^{i(\pi+4\pi)/5}, \quad e^{i(\pi+6\pi)/5}, \quad e^{i(\pi+8\pi)/5},$$

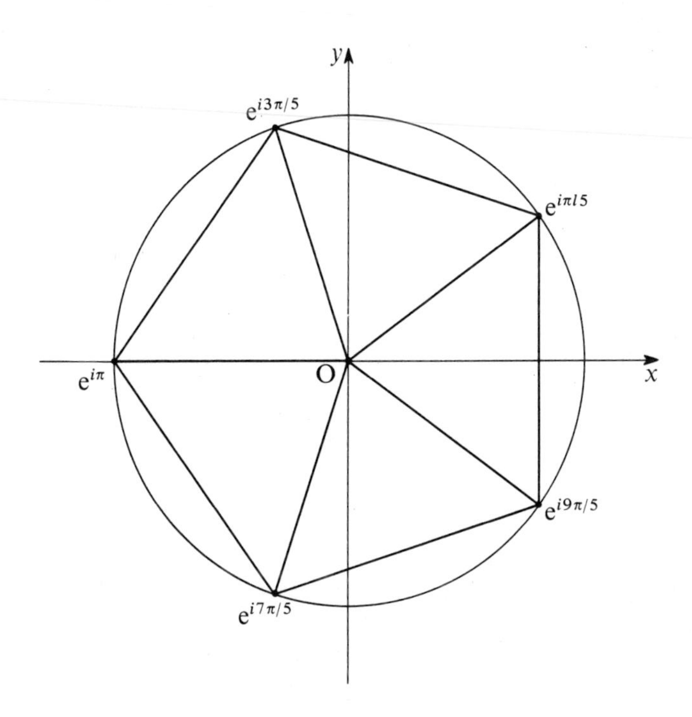

which when displayed on the unit circle lie at the vertices of a regular pentagon; as illustrated on the previous page.

2 Solve the complex equation $(1+z)^n = (1-z)^n$.

We require to solve the equation

$$\left(\frac{1+z}{1-z}\right)^n = 1,$$

which is equivalent to the given equation, since $z = 1$ is not a solution. Suppose $w = \dfrac{1+z}{1-z}$, so that we require the solutions of $w^n = 1$. Since $1 = \cos 2k\pi + i \sin 2k\pi$, i.e. $1 = \mathrm{e}^{i2k\pi}$, we have solutions

$$w = 1, \mathrm{e}^{i2\pi/n}, \mathrm{e}^{i4\pi/n}, \ldots, \mathrm{e}^{i2(n-1)\pi/n}.$$

If we denote a typical solution by $w = \mathrm{e}^{i\alpha}$, we then have

$$\frac{1+z}{1-z} = \mathrm{e}^{i\alpha},$$

so that

$$\begin{aligned} z &= \frac{\mathrm{e}^{i\alpha}-1}{\mathrm{e}^{i\alpha}+1} \\ &= \frac{\cos\alpha + i\sin\alpha - 1}{\cos\alpha + i\sin\alpha + 1} \\ &= \frac{2\sin\frac{\alpha}{2}\left(-\sin\frac{\alpha}{2} + i\cos\frac{\alpha}{2}\right)}{2\cos\frac{\alpha}{2}\left(\cos\frac{\alpha}{2} + i\sin\frac{\alpha}{2}\right)} \\ &= i\tan\frac{\alpha}{2}. \end{aligned}$$

The solutions of the original equation are therefore

$$z = i\tan\frac{\alpha}{2}, \text{ where } \alpha = 0, \frac{2\pi}{n}, \ldots, \frac{2(n-1)\pi}{n}.$$

Exercises 5.6

1 Find the square roots in $\mathbf{C}$ of $\sin 2\theta - i\cos 2\theta$, and $i-1$.

2 Solve the equation $x^5+2 = 0$ in $\mathbf{C}$.

3 Plot, in the real plane, the nth roots of -1, in case $n = 2, 3, 4$. Repeat for the corresponding nth roots of i.

Miscellaneous exercises 5

1 A complex number w is called a *primitive* nth *root of unity* if it is an nth root of 1 and corresponds to a rotation mapping ρ such that $I, \rho, \rho^2, \rho^3, \ldots, \rho^{n-1}$ are distinct, where I denotes the identity rotation $\rho_{(1,0)}$. Show that not all the roots of an equation $z^n = 1$ are primitive in case n is not a prime number.

If n is prime are all roots of 1 primitive?

2 Prove that, if w is a primitive nth root of unity, then given an integer r,

$$\begin{aligned} 1+w^r+w^{2r}+\ldots+w^{(n-1)r} &= 0 \text{ if } r \text{ is not a multiple of } n \\ &= n \text{ if } r \text{ is a multiple of } n. \end{aligned}$$

3 Prove that

$$x^{2n}-2x^n\cos\theta+1 = (x^n-e^{i\theta})(x^n-e^{-i\theta}).$$

Hence solve in the complex field the equation

$$x^{2n}-2x^n\cos\theta+1 = 0.$$

4 Compose the following functions

$$\begin{aligned} Z &= Z'+b && (b \text{ complex}) \\ Z' &= rz' && (r \text{ real} > 0) \\ z' &= e^{\theta i}z && (\theta \text{ real}) \end{aligned}$$

so as to obtain the function

$$Z = az+b$$

where $a = re^{\theta i}$. Deduce that the function $Z = az+b$ can be interpreted geometrically as a combination of a translation, a dilation, and a rotation of the co-ordinate plane.

5 Let z be a complex number such that $|z| = r \neq 0$ and $z = re^{i\theta}$. Prove that if

$$z\bar{z}+\bar{b}z+b\bar{z}+c = 0,$$

where c is real and b is such that $|b|^2 > c$, then $w = \frac{1}{r}e^{i\theta}$ is such that

$$cw\bar{w}+\bar{b}\bar{w}+bw+1 = 0.$$

Deduce that the transformation of complex numbers $Z = 1/z$ 'carries circles into circles, which may in particular cases be straight lines'. (The reader may wish to recall Worked examples 4.1, Example 3 and Exercises 1 4.1, Exercise 6.)

6 Prove that the composition of the transformations of **C** given by

$$Z = \frac{a}{c}+\frac{(bc-ad)Z'}{c}$$

$$Z' = \frac{1}{z'}$$

$$z' = cz+d$$

yields the 'bilinear' transformation

$$Z = \frac{az+b}{cz+d}.$$

Justify the statement 'bilinear transformations carry circles into circles, which may in particular cases be straight lines'.

7 If $z = x+yi$ is any complex number, with x and y real as usual, we define

$$e^z = e^x(\cos y + i \sin y),$$

where e^x denotes the real exponential function of the real variable x. Prove

(i) $e^{z_1}e^{z_2} = e^{z_1+z_2}$, for any complex numbers z_1, z_2,
(ii) $e^z e^{-z} = 1$, for any complex number z,
(iii) $e^z \neq 0$ for any complex number z,
(iv) if x is real, $|e^{ix}| = 1$,
(v) $e^z = 1$ if and only if $z = 2n\pi i$ for some integer n,
(vi) $e^{z_1} = e^{z_2}$ if and only if $z_1 - z_2 = 2n\pi i$ for some integer n,
(vii) if z is any complex number whose argument has principal value θ, then†

$$e^{(\log|z|+\theta i)} = z,$$

and any complex number w such that $e^w = z$ is of the form†

$$w = \log|z| + i(\theta + 2n\pi)$$

for some integer n.

8† If w is any complex number and $z = re^{i\theta}$, where $r = |z|$ and θ is the principal value of the argument of z, we define

$$z^w = e^{w(\log r + \theta i)}.$$

Prove $(1)^i = 1$; $(-1)^i = e^{-\pi}$; $(i)^i = e^{-\pi/2}$.

† The logarithms here are *natural*, i.e. to the base e.

APPENDIX ONE

General rules for associativity and commutativity

Whenever we have noted the two rules of associativity and commutativity in connection with an operation of either multiplication or addition of numbers we have always restricted ourselves to the simplest cases, namely $a(bc) = (ab)c$ and $ab = ba$ respectively, in multiplicative notation, or $a+(b+c) = (a+b)+c$ and $a+b = b+a$ respectively, in additive notation. Here we intend to show that these rules imply general rules for dealing with products or sums of more than two numbers.

First we should note that in what follows we use the multiplicative notation. It matters not which we use, additive or multiplicative, so we make this choice. Secondly we recall how we defined the product of more than two numbers. This was done inductively:

$$
\begin{aligned}
a_1a_2a_3 &= (a_1a_2)a_3,\\
a_1a_2a_3a_4 &= (a_1a_2a_3)a_4,\\
&\vdots\\
a_1a_2 \dots a_{n+1} &= (a_1a_2 \dots a_n)a_{n+1}.
\end{aligned}
$$

We now prove a result which enables us to clear any product of brackets, so that for example

$$(abc)((de)f) = abcdef.$$

Proposition. If $(ab)c = a(bc)$ for all a, b, c, then

$$(a_1 a_2 \dots a_m)(a_{m+1} \dots a_{m+n}) = a_1 a_2 \dots a_{m+n}.$$

Proof. We proceed by induction on n. Thus in case $n = 1$ the proposition states

$$(a_1 a_2 \dots a_m)(a_{m+1}) = a_1 a_2 \dots a_{m+1},$$

but this is precisely our definition of a product of $m+1$ terms. We therefore assume the result true if $n = k$ and deduce its truth for $n = k+1$:

$$\begin{aligned}
(a_1 a_2 \dots a_m)(a_{m+1} \dots a_{m+k+1}) \\
&= (a_1 a_2 \dots a_m)((a_{m+1} \dots a_{m+k})a_{m+k+1}) \\
&\quad \text{(by definition of } a_{m+1} \dots a_{m+k+1}) \\
&= ((a_1 a_2 \dots a_m)(a_{m+1} \dots a_{m+k}))a_{m+k+1} \\
&\quad \text{(by associativity rule: } a(bc) = (ab)c) \\
&= (a_1 a_2 \dots a_{m+k})a_{m+k+1} \\
&\quad \text{(by induction)} \\
&= a_1 a_2 \dots a_{m+k+1} \\
&\quad \text{(by definition of } a_1 a_2 \dots a_{m+k+1}).
\end{aligned}$$

As stated, this result allows us to clear any product of brackets. If we apply it to the particular example preceding the proposition we have

$$\begin{aligned}
(abc)((de)f) &= (abc)(def) \\
&= abcdef.
\end{aligned}$$

Having dealt with generalized associativity we next come to commutativity. For associativity we wished to show that any product is independent of the bracketing of the product, i.e. of the order in which the multiplications are to be carried out; for generalized commutativity we wish to prove that a product is independent of the order of the terms in it. For example, therefore, we wish to prove that

$$bdaefc = abcdef.$$

Proposition. If $ab = ba$ for all a, b and $a(bc) = (ab)c$ for all a, b, c, then given any permutation $(i_1 i_2 \ldots i_n)$ of $(1, 2, \ldots, n)$,

$$a_{i_1} a_{i_2} \ldots a_{i_n} = a_1 a_2 \ldots a_n.$$

Proof. Again we proceed by induction. For $n = 1$ the proposition is trivial (for $n = 2$ it simply restates the basic commutativity rule). We therefore assume it to be true for $n = k$ and deduce its truth for $n = k+1$. First let m be the integer such that $i_m = k+1$. Then

$$a_{i_1} a_{i_2} \ldots a_{i_{k+1}} = (a_{i_1} a_{i_2} \ldots a_{i_{m-1}})(a_{i_m}(a_{i_{m+1}} \ldots a_{i_{k+1}}))$$

(by associativity)

$$= (a_{i_1} a_{i_2} \ldots a_{i_{m-1}})(a_{i_{m+1}} \ldots a_{i_{k+1}})a_{i_m}$$

(by associativity and commutativity).

Now $(i_1, i_2, \ldots, i_{m-1}, i_{m+1}, \ldots, i_{k+1})$ is a permutation of $(1, 2, \ldots, k)$. Thus by associativity and induction

$$(a_{i_1} a_{i_2} \ldots a_{i_{m-1}})(a_{i_{m+1}} \ldots a_{i_{k+1}}) = a_{i_1} \ldots a_{i_{m-1}} a_{i_{m+1}} \ldots a_{i_{k+1}}$$

$$= a_1 a_2 \ldots a_k.$$

Since $i_m = k+1$, it follows that

$$a_{i_1} a_{i_2} \ldots a_{i_{k+1}} = (a_1 a_2 \ldots a_k) a_{k+1}$$

$$= a_1 a_2 \ldots a_{k+1},$$

as required.

Applied to our particular example preceding the proposition, we simply note that $bdaefc$ is a permutation of the letters $abcdef$, since each letter in the latter product occurs once and once only in the former; immediately then we deduce $bdaefc = abcdef$.

Exercises

Writing

$$a_1 + \ldots + a_m = (a_1 + \ldots + a_{m-1}) + a_m,$$
$$b_1 + \ldots + b_n = (b_1 + \ldots + b_{n-1}) + b_n,$$

use induction on m and n to prove that

$$(a_1 + \ldots + a_m)(b_1 + \ldots + b_n) = a_1 b_1 + \ldots + a_1 b_n + \ldots + a_m b_1 + \ldots + a_m b_n.$$

APPENDIX TWO

The existence of square roots in R

In this appendix we show how the completeness axiom[†] for the field **R** of real numbers can be used to prove the existence in **R** of unique positive square roots of positive real numbers. Apart from the all-important use of the axiom, the proof depends mainly on manipulation of suitably chosen inequalities.

Given a positive real number r we therefore consider the set X of all real numbers $x \geqslant 0$ and such that $x^2 \geqslant r$. Our aim will be to apply the completeness axiom to the set X to obtain its greatest lower bound, s say, and then to prove that s is such that $s^2 = r$. That X has indeed got a greatest lower bound will follow immediately from our axiom if we show that X is non-empty and bounded below. We therefore first prove two lemmas establishing these facts.

Lemma 1. The set X is non-empty.

Proof. If $r > 1$ then $r^2 > r$ and r is in X. If $r \leqslant 1$ then $r \leqslant 1^2$ and 1 is in X.

Lemma 2. The set X is bounded below.

[†] I.e. every non-empty set of real numbers which is bounded below has a greatest lower bound.

Proof. If $r \geqslant 1$, then $r \geqslant 1^2$, so for all x in X, $x^2 \geqslant 1^2$, i.e. $x \geqslant 1$, in other words, in this case X is bounded below by 1. On the other hand, if $r < 1$, then $r^2 < r$, i.e. for all x in X, $x^2 > r^2$, so $x > r$ and in this case X is bounded below by r.

As suggested, we now apply the completeness axiom, asserting that since X is non-empty and bounded below, there exists a greatest lower bound s for X. We further note that, since either 1 or r is a lower bound for X, s is greater or equal to 1 or r, both of which are positive, so $s > 0$.

There are now three possibilities. Either $s^2 < r$, $s^2 > r$ or $s^2 = r$. We assume first that $s^2 < r$ and second that $s^2 > r$; in both cases we deduce a contradiction, so that only the possibility $s^2 = r$ is left, and our proof is completed.

(i) If $s^2 > r$, consider the number x defined by

$$x = s - \frac{s^2 - r}{2s}.$$

We have

$$\begin{aligned} x^2 &= s^2 - (s^2 - r) + \left(\frac{s^2 - r}{2s}\right)^2 \\ &= r + \left(\frac{s^2 - r}{2s}\right)^2 \\ &> r. \end{aligned}$$

Since $x = \dfrac{s^2 + r}{2s}$, it is clearly positive and it follows that x is in X. But by definition $x < s$; so the assumption $s^2 > r$ implies that the greatest lower bound s of X is not a lower bound of X, which is a contradiction.

(ii) Suppose now that $s^2 < r$ and that δ is a positive real number to be fixed more precisely as we proceed. Since $\delta > 0$ and s is the greatest lower bound of X, it follows that $s + \delta$ is not a lower bound of X and there exists x in X such that $x < s + \delta$. But then

$$x^2 < s^2 + 2s\delta + \delta^2,$$

so if we choose $\delta \leqslant s$, as we surely may, then $\delta^2 \leqslant s\delta$ and we have

$$x^2 < s^2 + 3s\delta.$$

We now choose δ not only $\leqslant s$ but also such that $3s\delta \leqslant r - s^2$, e.g. we take δ to be the smaller of s and $\frac{r-s^2}{3s}$. We then have

$$\begin{aligned} x^2 &< s^2 + r - s^2 \\ &= r. \end{aligned}$$

But by definition $x^2 \geqslant r$, so the deduction $x^2 < r$ is a contradiction, again as required.

The uniqueness of s, which we asserted in the text, follows from the observation that s is positive and if $s^2 = r = t^2$, then $t = \pm s$. In other words there are two real square roots of r, one positive, one negative, namely $\pm\sqrt{r}$.

As we noticed to begin with, this proof depends essentially on the completeness axiom for the real numbers. We may also note that we have had to pick our inequalities fairly carefully and it is not at all obvious how to generalize the proof so as to prove the existence of nth roots for general n, or of solutions of equations such as $\sin x = r$ when $-1 < r < 1$, or other similar results. What is needed is machinery which will produce all these facts. Such machinery is part of the subject of analysis. But that is another story.

Index